11th International Conference on Environmental Science and Technology (ICEST 2020)

Held online due to COVID-19

IOP Conference Series: Earth and Environmental Science
Volume 728

Beijing, China
19 – 21 November 2020

ISBN: 978-1-7138-2947-8
ISSN: 1755-1307

Printed from e-media with permission by:

Curran Associates, Inc.
57 Morehouse Lane
Red Hook, NY 12571

Some format issues inherent in the e-media version may also appear in this print version.

Institute of Physics
Dirac House, Temple Back
Bristol BS1 6BE UK

Phone: 44 1 17 929 7481
Fax: 44 1 17 920 0979

techtracking@iop.org

Additional copies of this publication are available from:

Curran Associates, Inc.
57 Morehouse Lane
Red Hook, NY 12571 USA
Phone: 845-758-0400
Fax: 845-758-2633
Email: curran@proceedings.com
Web: www.proceedings.com

TABLE OF CONTENTS

HYDROLOGY AND WATER RESOURCES MANAGEMENT

SOLID WASTE TREATMENT AND SOIL POLLUTION ASSESSMENT

ENERGY SAVING AND EMISSION REDUCTION

INDUSTRIAL PRODUCTION STANDARDS AND ENVIRONMENTAL IMPACT

IOP Conf. Series: Earth and Environmental Science **728** (2021) 011001 doi:10.1088/1755-1315/728/1/011001

PREFACE

The 2020 11th International Conference on Environmental Science and Technology (ICEST 2020) was held as virtual conference during November 19-21, 2020 due to recent pandemic. The decision to hold the virtual conference was made by conference committees in compliance with many travel restrictions and regulations that were imposed by countries all over the world. The health and safety of our participants and members is top priority to the conference organization. Therefore, worldwide participants have attended this online conference by zoom application and finished their oral presentations successfully.

Conducted by the organizing committees in Beijing, there are around 70 participants in total and they are from UK, Indonesia, Australia, Canada, Japan, Romania, China, Thailand and so on. ICEST 2020 is supported by Beijing Normal University (BNU), Xiamen University of Technology (XMUT), International Society for Environmental Information Sciences (ISEIS), South-South Collaborative and Sustainable Development Center (SSCSDC), and Fujian Smart City Association (FSCA). This three-day conference focused on the research fields in environmental science and technology. The primary goal of the conference is to promote research and development activities in environmental science and technology. Another goal is to promote scientific information interchange among researchers, developers, engineers, students, and practitioners working in China and abroad.

During the conference, the conference model was divided into mainly two parts, including keynote speakers and oral presentations. In the first part, we invited internationally well-known experts to deliver speeches as keynote speakers. They are Prof. Yongping Li from Beijing Normal University, China; Assoc. Prof. Farhad Shahnia from Murdoch University, Australia; Prof. Guilin Zheng from Wuhan University, China; Prof. Wei-Jen Lee from University of Texas at Arlington, USA; Assoc. Prof. S. M. Muyeen from Curtin University, Australia; Prof. Zhijun Peng from University of Bedfordshire, UK. In the second part, each scholar was given 12 minutes to perform their oral presentation and 3 minutes for Q&A. All papers presented at the 2020 11th International Conference on Environmental Science and Technology (ICEST 2020) are included in this volume, which contains four chapters with topics: (1) Hydrology and Water Resources Management, (2) Solid Waste Treatment and Soil Pollution Assessment, (3) Energy Saving and Emission Reduction, (4) Building Climatology and Bioclimatology. All papers are subjected to peer-review by conference committee members and international reviewers. The papers are selected based on high quality and high relevancy to the conference scope.

We would like to express our gratitude to all individuals and organizations who supported the ICEST 2020. Their helps and contributions are of great importance to the success of this conference. In addition, we would like to thank the organizing committee for its valuable inputs in shaping the conference program and reviewing the submitted papers. We sincerely hope that the ICEST 2020 turned out to be a forum for excellent discussions that enabled new ideas to come about and promoted collaborative research works. We are sure that the proceedings will serve as an important research source of references and knowledge, which will lead to not only scientific and engineering findings but also new products and technologies.

Yongping Li
Beijing Normal University, China
Dec. 16, 2020

Committees

Conference Chairs

Gordon Huang, International Society for Environmental Information Sciences, Canada
Yongping Li, Beijing Normal University, China

Technical Program Committee

James T. Anderson, West Virginia University, USA
Guangwei Huang, Sophia University, Tokyo, Japan
Lixiao Zhang, Beijing Normal University, China

Guest Editors

Henrique J. O. Pinho, Instituto Politécnico de Tomar, Portugal
Dina M. R. Mateus, Instituto Politécnico de Tomar, Portugal

Organizing Committee

Hong Kong Chemical, Biological & Environmental Engineering Society, Hong Kong

Contact Chair

Ms Erica Han, Hong Kong Chemical, Biological & Environmental Engineering Society, Hong Kong

Technical Members

Liang Fang, Jiangping Bio-Substrate, China
Adam Fenech, University of Prince Edward Island, Canada
Vasilis Fthenakis, Columbia University, USA
Tae-Hyoung Gim, Seoul National University, South Korea
Radu Godina, NOVA University Lisbon, Portugal
Yan Hao, Beijing Normal University, China
Ryusuke Hashimura, International Institute of Natural Disaster, Japan
Wendy Huang, McMaster University, Canada
Yee-Chang Jin, University of Regina, Canada
Vuttichai Kaewkrajang, Pollution Control Department, Thailand
Monzur Imteaz, Swinburne University of Technology, Australia
Zoe Li, McMaster University, Canada
Chuxia Lin, University of Salford, UK
Jinliang Liu, Ontario Ministry of Environment and Climate Change, Canada
Minghua Liu, Fuzhou University, China
Reza Nekovei, Texas A&M University-Kingsville, USA
Kriansak Pirarai, Department of Groundwater Resource, Thailand
Makridis Sofoklis, University of Patras, Greece
Vladimir Strezov, Macquarie University, Australia
Lerdphan Sukyirun, Office of the National Water Resources, Thailand

IOP Conf. Series: Earth and Environmental Science **728** (2021) 011002 doi:10.1088/1755-1315/728/1/011002

Peer review declaration

All papers published in this volume of IOP Conference Series: Earth and Environmental Science have been peer reviewed through processes administered by the Editors. Reviews were conducted by expert referees to the professional and scientific standards expected of a proceedings journal published by IOP Publishing.

- **Type of peer review:**
  Double-blind
- **Conference submission management system:**
  Iconf submission system
- **Number of submissions received:**
  26 papers
- **Number of submissions sent for review:**
  26 papers
- **Number of submissions accepted:**
  14 papers
- **Acceptance Rate (Number of Submissions Accepted / Number of Submissions Received X 100):**
  53.84%
- **Average number of reviews per paper:**
  2 papers
- **Total number of reviewers involved:**
  13 reviewers
- **Any additional info on review process:**

1. Preliminary Review
   When we received authors' paper(abstract and full paper), we will submit the paper to auditing department for checking, the auditing department will review the content, format and grammers.
2. The full papers passed the first review will be eviewed again by conference technical committees from the following aspects: Originality, Innovation, Technical Soundness, Applicability, Clarity of presentation and Relevance.
3. Only the paper passed the first review and the second review, it can be published.

- **Contact person for queries:**
  Prof. Yongping Li
  Beijing Normal University, China
  yongpingli33@163.com

IOP Conf. Series: Earth and Environmental Science **728** (2021) 012001 doi:10.1088/1755-1315/728/1/012001

Rainfall Threshold of Citarum River Flood for Early Warning on Urban Area

Ariani Budi Safarina[1], Chairunnisa[1], Novita Dwi Lestari[1], Devi Aprilianingsih[1], Fathulhuda Muhammad Syahrullah[2], Muthiah Sadidah[3]

1 Civil Engineering Department, Universitas Jenderal Achmad Yani, Jalan Terusan Jenderal Sudirman PO BOX 148, Cimahi, 40511, Indonesia
2 Geodesy and Geomatics Engineering Department, Institut Teknologi Bandung, Jalan Ganesha 10, Bandung, 40116, Indonesia
3 Civil Engineering Department, Institut Teknologi Bandung, Jalan Ganesha 10, Bandung, 40116, Indonesia
E-mail: ariani.budi@lecture.unjani.ac.id

Abstract. Floods in Bandung basin located in the upper Citarum river resulted in huge losses because they occurred in industrial areas and densely populated settlements. Disaster risk reduction can be done with flood early warning in residential areas. The purpose of this study is to determine the rainfall threshold that causes inundation in residential areas in seven districts located on the floodplain of the Citarum river. The method used is the analysis of the rainfall-runoff model to obtain the Citarum river unit hydrograph. Flows in the rivers are assumed to be one-dimensional unsteady flow whereas overflow is assumed to be overland flow. Threshold rainfall is determined based on the deconvolution method, with the drainage coefficient calculated based on the land use map. The results of the study are the capacity discharge and the load discharge of the Citarum river in the seven sub-districts, which the most upstream is the Majalaya sub-district and the downstream is Katapang sub-district. The lowest threshold rainfall has occurred in the sub-district Bojongsoang that is 54.34 mm that produced the flood water level as high as 662 m MSL. and the evacuation time for early warning is 5.5 hours.

1. Introduction

The Citarum River Basin in West Java Province is a National Strategic Watershed that has very large potential water resources, including the potential of energy, those are water sources of 3 power plant reservoirs as the electricity suppliers for Java and Bali, food potential, those are water sources of the irrigation network, and potential raw water for the community and industry. The current problem is the destruction of the Citarum watershed because of the greater runoff causing flooding. Bandung Regency which is included in the Bandung Basin is an area that experiences Citarum River overflow every year. Floods in the Bandung Basin region occur because of reduced water infiltration into the ground so that each time the rain produces a large proportion of runoff water and accumulates into flooding and inundation [1].

Settlements along the Citarum River Basin are very vulnerable to flooding. In a short time, the floods soaked the settlement area and stopped and paralyzed the socio-economic activities of the surrounding area. Flooding continues even though local rains have subsided but rain in other regions continues to occur because the flow of rain in other locations still leads to the Citarum River through tributaries. Floods due to overflowing of the Citarum River are becoming more frequent, with areas inundated not only along settlements in river basins but already expanding into the surrounding areas along the

Citarum River Basin [2]. Changes in the extent of flooding in South Bandung are currently very severe, not only the increase in the area but also the increase in flood water level. The floodwater level at Cieunteung Village, Bale Endah Village, Bale Endah District, Bandung Regency is reported to exceed 4 meters. One reason is the change in land use in the upper reaches of the Citarum River and in the Bandung Basin, which is an urban and industrial area. In 2010, several textile industries stopped production due to storm rainfall which causes flooding [3].

Various efforts have been made by the government to prevent residents affected by flooding. Technically, the Bandung regency government supported by the West Java Provincial Government and the Central Government has normalized the Citarum river. The shallow riverbed is then dredged to restore the river's depth so that the surface of the river is expected not to overflow. Social handling has also been carried out, such as evacuation and relocation of permanent resettlement of residents to avoid overflowing river water. But these various efforts in their realization are not entirely optimal in preventing people from flooding. One of the main obstacles is believed to originate from the condition of the community itself. Increased population, overcrowding, and attitudes the community is believed to have relevance to the success of the flood disaster in the government programs to deal with the flood problem. One of the adaptation strategies to deal with floods is to form groups in the community, especially in flood-prone areas that aim to deal with flood disasters. The emergence of socially conscious communities indicates the application of survival values and overcoming problems that are based on the independence of the community itself [4].

The early warning system gives people time to avoid floods, tornadoes, or tsunamis so that local stakeholders can evacuate and protect many people. This system provides information about events that endanger public health so that the response to aid can be faster such as food needs and feelings of insecurity. Early warnings given can also provide an opportunity for the community to protect its property and infrastructure [5]. The parameter used as an input for a community-based flood warning system is the elevation of the river's water level which is monitored by people living near the river [6][7]. If the river water level reaches a certain elevation that has been designated as a flood elevation, then an early warning alarm will be reads then this information is disseminated to stakeholders through a decision-supporting system [8]. The flood early warning system can also use rainfall data as a threshold by reading rainfall data from the rain station [9]. Rainfall observation systems are also used in flood early warning systems in Kenya Africa [10].

2. Methods of the study

The method used is the analysis of the rainfall-runoff model to obtain the Citarum river unit hydrograph. Flows in the rivers are assumed to be one-dimensional unsteady flow whereas overflow is assumed to be overland flow [11]. Threshold rainfall is determined based on the deconvolution method, with the drainage coefficient calculated based on the land use map.

In the rainfall-runoff model analysis, the unit hydrograph is calculated based on measured data or the observation unit hydrograph. The observation unit hydrograph method used in this study is the convolution method.

An impulse, both a step and a pulse response function, is defined as having a continuous-time domain. If the time domain is discretized with a duration interval Δt, then there are two ways to describe the continuous-time function in the discrete-time domain, the pulse data system, and the sample data system. The pulse data system is used for precipitation and the sample data system is used for direct flow and runoff. The discrete convolution equation for a linear system is,

$$Qn = \sum_{m=1}^{n \leq m} P_m U_{n-m+1} \qquad (1)$$

where Q is direct runoff discharge, P is effective rainfall and U is unit hydrograph ordinate. The basic flow is determined using the straight-line method of flow hydrograph. The synthetic unit hydrograph derived from Nakayasu method, while the observation unit hydrograph (U) is obtained by deconvolution calculation as follows,

$$U = \{(P^T) . P\}^{-1} . P^T . Q \qquad (2)$$

The equation for one-dimensional unsteady flow used in this study can be seen in the following equation,

$$\frac{\partial y}{\partial x} + \alpha \frac{V}{g}\frac{\partial V}{\partial x} + \frac{1}{g}\frac{\partial V}{\partial t} + \frac{\partial z}{\partial x} + S_f = 0 \qquad (3)$$

where y is the flow depth, V is velocity, z is the distance of the bed channel from the datum, and S_f is energy slope. Unsteady flow analysis can identify the maximum water surface elevation at various cross-sections during the propagation of the flood wave through a river reach. The unsteady flow in a river which meanders through a flood plain is complicated because of differences in hydraulic resistances of the main river channel and the flood plain and variation in the cross-sectional geometries of the channel [12].
The overflows from the river when it floods and flows to the land, in this study is assumed to be a turbulent overland flow with the direction from the river to the land [13]. The equation used to determine the inundation area is based on land elevation and the inundation height (Y) equation for overland flow as follows,

$$Y = \left(\frac{nq_o}{S_o^{\frac{1}{2}}}\right)^{3/5} \qquad (4)$$

where n is the roughness coefficient, q_o is the discharge per unit width and S_o is the bed slope.
Determine the monthly flood discharge for each sub-district is done by analyzing the rainfall-runoff model in the sub-district for analysis of inundation in the respective sub-district. Overflow to the land is used to calculate flood elevation for each sub-district and threshold rainfall for each sub-district is calculated using the convolution method.

3. Results and discussion

3.1. Study area

The upstream Citarum river basin where the outlet is in Nanjung, Margaasih Sub-District at an elevation of 675 m MSL as shown in figure 1, is located at -6,75^0 S to -7,24^0 S and 107,37^0 E to 107,94^0 E with an area of about 1762 km^2 and river length 82 km. The upstream of the Citarum river is in Kertasari Sub-District at an elevation of 1638 m MSL.
The flow of the Upper Citarum watershed crosses 4 cities including Bandung Regency, West Bandung Regency, Bandung City, and Cimahi City. In addition to being a source of raw water for drinking water, the Citarum River also being an irrigation water source for hundreds of thousands of hectares of rice fields and power plants for Java and Bali. In the Citarum river, there are three large reservoirs, namely Saguling, Cirata, and Jatiluhur. The map of Upper Citarum watershed is shown in figure 1,
The upstream area of the Citarum river starts from Majalaya Sub-district to Margaasih Sub-district, geographically shaped in a basin so it is often called the Bandung Basin. The Bandung Basin is a bowl-shaped at an elevation of 662.5 m MSL and in a hydrological model is a water parking area so it is prone to flooding. The Bandung Basin area is surrounded by mountains starting from Manglayang Mountain at an elevation of 1800 m MSL in Cilengkrang Sub-District, Geulis Mountain in Tanjungsari Sub-District at an elevation of 1200 m MSL, Kerenceng Mountain in Cimanggung Sub-District at an elevation of 1690 m MSL, Mandalawangi Mountain in Nagreg Sub-District at an elevation of 1600 m MSL, Artapela Mountain in Pangalengan Sub-District at an altitude of 1525 m MSL, Bumi Wangi Mountain in Ciparay Sub-District at an elevation of 1027 m MSL and Pasir Ipis Mountain in Cililin Sub-District at an elevation of 1050 m MSL.

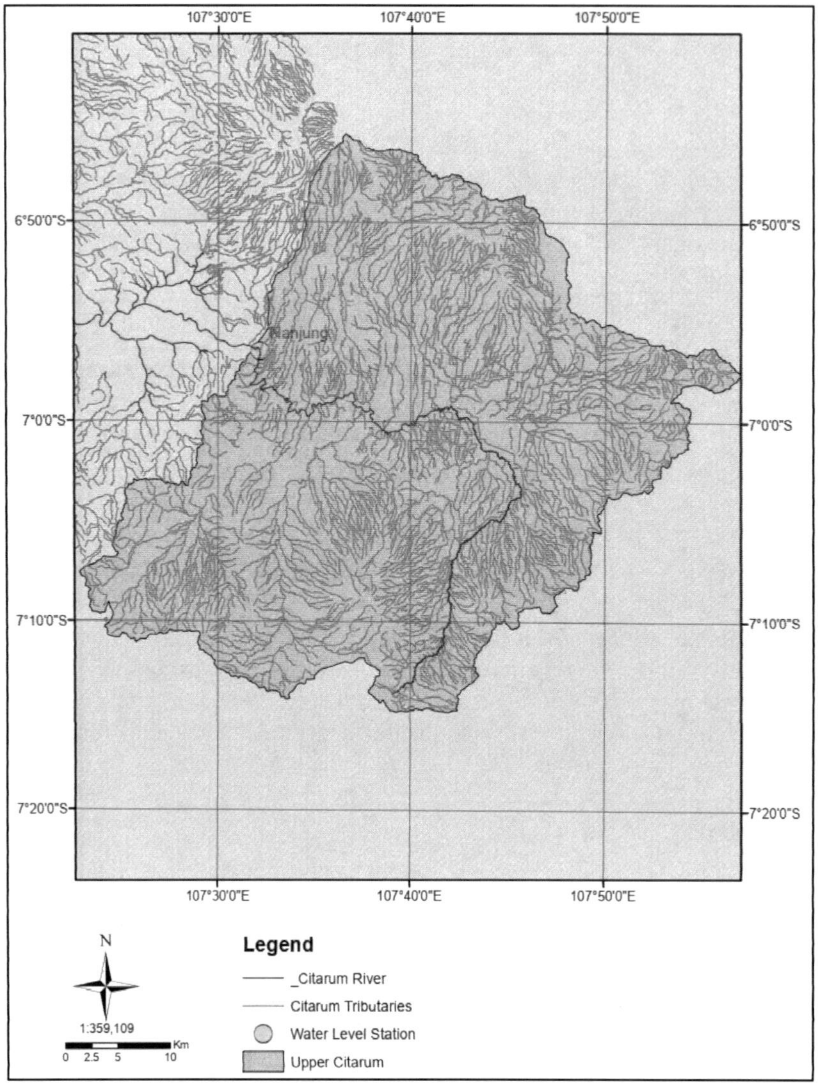

Figure 1. Upper Citarum River Basin Map

Based on the land use map, the runoff coefficient of the Upper Citarum river basin is 0.57.

3.2. Citarum river unit hydrographs in each sub-district

Subdistricts on the banks of the upstream Citarum river that is often flooded areas analyzed in this study are Majalaya Sub-District, Solokan Jeruk Sub-District, Ciparay Sub-District, Bojongsoang Sub-District, Baleendah Sub-District, Dayeuhkolot Sub-District, and Katapang Sub-District. To analyze the overflow that occurred, the Citarum river unit hydrograph was derived in the river segments through these seven districts. To get the unit hydrograph, the characteristics of the river in the section are measured, namely the length of the main river and the area of the watershed. The unit hydrograph is calculated using the Nakayasu synthetic method. Peak discharge in the unit hydrograph if multiplied by convolution with effective rainfall shows the maximum discharge load on that cross-section, while peak time indicates the time of evacuation in disaster mitigation.

The monthly design rainfall is calculated based on a 50 year return period with the frequency analysis method from Gumbel. Based on this rainfall, the discharge load calculated for each sub-district is also a monthly discharge. This monthly analysis is done so that flood observations can be analyzed in more detail with the behavior of the seasons. Indonesia is a tropical country with two seasons. The peak wet season generally occurs in November, December, and January, while the peak dry season in June, July, and August. The table 1 shows unit hydrograph parameters of the Citarum river in each sub-district,

Table 1. Unit Hydrograph Parameters of Citarum River in Each Sub-District

Sub-District Number	Sub-District Name	Unit Hydrograph Parameters		
		Qp (m^3/s)	Tp (hour)	Tb (hour)
1	Majalaya	6.04	3.69	35.00
2	Solokan Jeruk	5.67	5.00	40.00
3	Ciparay	5.96	4.65	40.00
4	Bojongsoang	12.57	5.52	50.00
5	Baleendah	14.24	6.00	55.00
6	Dayeuhkolot	16.78	6.00	65.00
7	Katapang	19.23	7.60	75.00

Based on the table 1, the highest unit hydrograph parameter is at the farthest location from the upstream of the Citarum river, Katapang. This means that the discharge load from the Citarum river channel gets greater if the location is further away from the upstream. Figure 2 shows a map of sub-districts along the banks of the upper Citarum river which are the location of the study in this research.

Figure 2. Map of Sub-Districts Along the Banks of the Upper Citarum River

3.3. Inundation area in each sub-district

Determine the inundation area for each sub-district using the elevation of flood water levels obtained from the calculation of overland flow based on overflow discharge. One-dimensional unsteady river analysis was performed using the Hecras software. Inputs entered into the software are river cross-sections and flow hydrographs. Flow hydrographs are obtained by convolution multiplying between the effective rainfall and the unit hydrograph. Table of monthly calculation results for each sub-district is shown in table 2,

Table 2. Monthly Inundation Characteristics in Each Sub-District

Sub-District Number	Sub-District Name	January	February	Marh	April	Mei	June	July	August	Sept	Oct	Nov	Dec
1	**Majalaya**												
	Qload (m^3/s)	193.06	348.60	284.46	309.54	309.54	255.20	253.58	226.25	306.36	330.08	324.09	339.84
	Qcap (m^3/s)	242.50	242.50	242.50	242.50	242.50	242.50	242.50	242.50	242.50	242.50	242.50	242.50
	Flood Elevation (m MSL)	No Flood	674.45	673.96	674.17	674.17	673.62	673.60	No Flood	674.15	674.32	674.28	674.39
	Inundation Area (km^2)	No Flood	4.57	2.62	3.47	3.47	1.28	1.18	No Flood	3.37	4.07	3.90	4.34
	% inundation area	No Flood	18.92	10.85	14.37	14.37	5.30	4.88	No Flood	13.95	16.87	16.16	17.97
2	**Solokan Jeruk**												
	Qload (m^3/s)	195.61	288.54	297.84	245.82	259.97	238.28	240.52	185.67	266.67	293.29	312.35	342.80
	Qcap (m^3/s)	242.50	242.50	242.50	242.50	242.50	242.50	242.50	242.50	242.50	242.50	242.50	242.50
	Flood Elevation (m MSL)	No Flood	674.06	674.15	673.45	673.72	No Flood	No Flood	No Flood	673.81	674.10	674.27	674.51
	Inundation Area (km^2)	No Flood	14.06	15.70	2.90	7.86	No Flood	No Flood	No Flood	9.55	14.91	18.05	22.43
	% inundation area	No Flood	58.21	65.01	12.01	32.54	No Flood	No Flood	No Flood	39.55	61.74	74.76	92.88
3	**Ciparay**												
	Qload (m^3/s)	384.00	312.57	403.89	391.67	363.99	405.53	224.88	145.24	331.02	338.55	530.67	365.73
	Qcap (m^3/s)	275.70	275.70	275.70	275.70	275.70	275.70	275.70	275.70	275.70	275.70	275.70	275.70
	Flood Elevation (m MSL)	666.88	666.37	667.00	666.93	666.76	667.01	No Flood	No Flood	666.52	666.58	667.61	666.77
	Inundation Area (km^2)	6.90	3.61	7.63	7.19	6.10	7.69	No Flood	No Flood	4.61	4.98	11.53	6.17
	% inundation area	12.73	6.67	14.08	13.26	11.26	14.19	No Flood	No Flood	8.50	9.18	21.27	11.39
4	**Bojongsoang**												
	Qload (m^3/s)	583.35	531.51	815.60	576.30	482.62	458.94	535.05	239.17	493.36	434.07	533.04	694.27
	Qcap (m^3/s)	389.35	389.35	389.35	389.35	389.35	389.35	389.35	389.35	389.35	389.35	389.35	389.35
	Flood Elevation (m MSL)	663.45	663.20	664.33	663.42	662.94	662.78	663.22	No Flood	663.00	662.60	663.21	663.90
	Inundation Area (km^2)	16.13	13.38	25.86	15.77	10.39	8.72	13.58	No Flood	11.09	6.69	13.47	21.15
	% inundation area	57.49	47.70	92.19	56.23	37.05	31.08	48.41	No Flood	39.55	23.83	48.01	75.41
5	**Baleendah**												
	Qload (m^3/s)	917.48	746.82	964.99	935.80	869.66	968.92	537.29	347.02	790.90	808.90	1267.92	873.82
	Qcap (m^3/s)	514.83	514.83	514.83	514.83	514.83	514.83	514.83	514.83	514.83	514.83	514.83	514.83
	Flood Elevation (m MSL)	664.11	663.37	664.29	664.18	663.92	664.30	661.96	No Flood	663.58	663.66	665.30	663.93
	Inundation Area (km^2)	11.95	8.58	12.78	12.27	11.08	12.84	2.11	No Flood	9.53	9.90	17.40	11.16
	% inundation area	29.82	21.42	31.88	30.62	27.64	32.05	5.28	No Flood	23.77	24.69	43.41	27.83
6	**Dayeuhkolot**												
	Qload (m^3/s)	924.85	656.33	843.05	683.84	537.57	565.10	637.93	547.54	1001.95	614.46	848.67	872.50
	Qcap (m^3/s)	609.10	609.10	609.10	609.10	609.10	609.10	609.10	609.10	609.10	609.10	609.10	609.10
	Flood Elevation (m MSL)	662.54	661.42	662.27	661.59	No Flood	No Flood	661.29	No Flood	662.77	661.04	662.29	662.37
	Inundation Area (km^2)	4.92	1.57	4.11	2.07	No Flood	No Flood	1.17	No Flood	5.61	0.43	4.17	4.41
	% inundation area	44.33	14.18	37.03	18.67	No Flood	No Flood	10.54	No Flood	50.54	3.84	37.56	39.76
7	**Katapang**												
	Qload (m^3/s)	364.08	855.36	573.54	427.19	606.23	469.99	404.51	240.49	472.89	560.47	848.33	659.33
	Qcap (m^3/s)	609.10	609.10	609.10	609.10	609.10	609.10	609.10	609.10	609.10	609.10	609.10	609.10
	Flood Elevation (m MSL)	No Flood	660.41	No Flood	No Flood	No Flood	No Flood	No Flood	No Flood	No Flood	No Flood	660.39	659.54
	Inundation Area (km^2)	No Flood	0.81	No Flood	No Flood	No Flood	No Flood	No Flood	No Flood	No Flood	No Flood	0.80	0.31
	% inundation area	No Flood	5.12	No Flood	No Flood	No Flood	No Flood	No Flood	No Flood	No Flood	No Flood	5.04	1.97

Based on Table 2 above, November and December are months that always occur flooding in each sub-district, while February is the month that does not experience. The highest inundation occurred in December in Solokan Jeruk and in March in Bojongsoang with a pool area of 92.2% of the subdistrict's area. Bojongsoang and Baleendah sub-districts always experience floods throughout the year, except in August. Katapang sub-district is the sub-district with the least frequency of flood events, those are only in March, November, and December.

3.4. Rainfall threshold in each sub-district

Threshold rainfall is calculated based on river discharge capacity in each district. This rainfall can be seen from the rainfall gauge that is published on the rainfall station display so that it is easily accessible both by the community and by stakeholders. The rainfall stations used for the analysis of each district differ according to station location. Representative rain stations are in the upstream sub-districts. The rain station which is located under the district certainly does not affect the discharge analyzed. Rainfall station for each sub-district is shown in table 3,

Table 3. Rainfall Threshold in Each Sub-District

Sub-District Number	Sub-District Name	Rainfall Threshold		
		R_{eff} (mm)	Gauge Rainfall (mm)	Evacuation Time (hour)
1	Majalaya	40.15	70.44	3.69
2	Solokan Jeruk	40.69	71.38	5.00
3	Ciparay	46.26	81.16	4.65
4	Bojongsoang	30.97	54.34	5.52
5	Baleendah	36.15	63.43	6.00
6	Dayeuhkolot	36.30	63.68	6.00
7	Katapang	31.67	55.57	7.60

Based on Table 3, each sub-district has a different rainfall threshold and evacuation time. This difference shows that each sub-district has different characteristics in flood events [14]. This result can be detailed information for each sub-district for flood disaster mitigation. Overall, the threshold rainfall used was the lowest, so for the Bandung basin flooding the threshold rainfall is the threshold rainfall in Bojongsoang which is 54.34 mm and the evacuation time is 5.52 hours, with the water level elevation at 662 m MSL.

4. Conclusion

Urban flooding is different from river basin flooding because it requires a detailed analysis of overland flow in urban areas for controlling the flood. The overland flow analysis that is integrated with flood analysis in watersheds, becomes an integrated solution between watershed analysis and urban flood analysis. Threshold rainfall in each sub-district can provide flood alert information in more detail because floods that occur in the most vulnerable sub-districts will occur first so that a warning will be given at the earliest. In this study, Bojongsoang is the most vulnerable district with the lowest threshold rainfall of 54.34 mm.

5. References

[1] R Maria, W Naily, A Purwoarminta, Hendarmawan. 2019. Artificial Recharge as an Effort to Increase Urban Water Resilience. IOP Conf.Series: Earth and Environmental Science **306** 012018, pp. 1-11

[2] Aton Patonah, Edi Tri hayanto, Pradnya P Raditya Rendra, Murni Sulastri, Zulfa Sabily, Endah Tresnasari, Emi Sukiyah, Panji Ridwan, Muhammad Jihad Abdurrohman, Agam Subagja, and Alvindo A.Putra. 2019. Flood Potential in The Downstream Of Citarum River, Muara Gembong, Bekasi District, West Java. Journal of Geological Sciences and Applied Geology, Vol 3(3): p. 29-35

[3] Dedi Mulyadi, Igna Hadi S, Andarta F. Khoir. 2010. South Bandung Flood in 1985 – 2010 Based on Landuse Change and Rainfall Fluctuation (Banjir Bandung Selatan Tahun 1985 – 2010 Ditinjau dari Perubahan Tutupan lahan dan Fluktuasi Curah Hujan). Presentation of Research Results of Geotechnology Research Center – LIPI Proceeding, 241-250

[4] Tachya Muhamad, Bintarsih Sekarningrum, Yusar. 2017. Social Capital in Flood Disaster Management (Case in Bandung Regency, West Java). Sosioglobal Journal 1(2):101-114

[5] Mohamad Sukeri Bin Khalid, Shazwani Binti Shafiai. 2015. Flood Disaster Management in Malaysia: An Evaluation of the Effectiveness Flood Delivery System, International Journal of Social and Humanity, Vol.5 No.4 p.398-402

[6] Rokon Uddin Md, M.Abdul Awal. 2013, Early Warning on Disastrous Weather through Cell Phone, IOSR Journal of Computer Engineering, Volum 11 Issue 5 p.51-55

[7] Fakhruddin S.H.M, Akiyuki Kawasaki, Mukand S Babel. 2015. Community Responses to Flood Early Warning System: Case Study in Kaijuri Union, Bangladesh, International Journal of Disaster Risk Reduction- Elsevier, p.1-9

IOP Conf. Series: Earth and Environmental Science **728** (2021) 012001 doi:10.1088/1755-1315/728/1/012001

[8] Keoduangsine Saysoth, Robert Goodwin. 2012. An Appropriate Flood Warning System in the Context of Developing Countries, International Journal of Innovation, Management and Technology, Vol.3 N0.3,p.213-216

[9] Ariani Budi Safarina, Ramli, Muhammad Shiddiq Sayyid Hasharon. 2017. Two Alert Flood Early Warning System Method Based on Rainfall-Runoff Model International Journal of Civil Engineering and Technology 8(12): 1032-1044

[10] Hoedjes Joost C.B, Andrekooiman, Ben H.P. Maathuis, Mohammed Y.Said, Robert Becht, Agnes Limo, Mark Mumo, Joseph Nduhiu-Mathenge, Ayub Shaka, Bob Shu. 2014. A Conceptual Flash Flood Early Warning System for Africa, Based on Terrestrial Microwave Links and Flash Flood Guidance, International Journal of Geo-Information, Issue 3, p.584-598

[11] Ariani Budi Safarina, Hang Tuah Salim, Iwan K Hadihardaja, M.Syahril BK. 2011. Clusterization of Synthetic Unit Hydrograph Methods Based on Watershed Characteristics, International Journal of Civil and Environmental Engineering, Vol.11 No.6 p.76-85

[12] Ariani Budi Safarina, Iin Karnisah, Ade Sena Permana. 2019. Study of The Effect of River Basin Morphology Change on Threshold Parameters in Cimahi Flood Early Warning System. IOP Conf.Series: Earth and Environmental Science 273 012013, pp. 1-8

[13] Ariani Budi Safarina, Iin Karnisah, Ade Sena Permana, Dadang Sufianto. 2019. Green Cimahi Watershed for Balancing Water Supply and Flood Control. IOP Conf.Series: Materials Science and Engineering 669, 012040, pp. 1-11

[14] Anwar, M. R. (2011). The Rainfall-Runoff Model Using of The Watershed Physical Characteristics Approach. Journal of Civil and Environmental Engineering, 71-75

Acknowledgments

This research is fully supported by UNJANI Competitive Research SKEP/129/UNJANI/V/2020. The authors fully acknowledged Jenderal Achmad Yani University for the approved fund which makes this important research viable and effective.

IOP Conf. Series: Earth and Environmental Science **728** (2021) 012002 doi:10.1088/1755-1315/728/1/012002

Classification of Land Cover in Kumamoto Area, Japan to Evaluate Infiltration and Surface Runoff

H Amano[1], Y Iwasaki[2] and T Ichikawa[1]

1 Liberal Arts Education Center, Kumamoto Campus, Tokai University, 9-1-1 Toroku, Higashi-ku, Kumamoto 862-8652, Japan
2 Department of Electrical Engineering and Computer Science, Faculty of Industrial and Welfare Engineering, Tokai University, 9-1-1 Toroku, Higashi-ku, Kumamoto 862-8652, Japan
E-mail: ah069881@tsc.u-tokai.ac.jp

Abstract. Almost 100% of domestic water is covered by groundwater in the Kumamoto region, Japan. However, in recent years, the groundwater recharge area (particularly paddy fields) has been significantly decreasing, and the groundwater level has been gradually dropping in some places. Aside from this, the groundwater level was abnormally changed as an impact of the 2016 Kumamoto earthquake. Our final goal is to build a model to represent the groundwater level change including the abnormal change. For this purpose, we generated a land cover map by using temporal Normalized Difference Vegetation Index (NDVI) changes based on Landsat-8 satellite images and Geographic Information System (GIS) data. Our classification results well expressed the actual land cover especially around rivers, lakes and buildings compared with the existing land cover map. The calculated infiltration and surface runoff amounts from grasslands with the different land uses indicated the importance of understanding not only land covers but also land uses to evaluate infiltration and surface runoff amounts.

1. Introduction

Kumamoto region in Japan has been using groundwater for the public water of almost 100 %. Therefore, conserving groundwater is crucial from the viewpoint of both quantity and quality. In recent years, the nitrate pollution mainly due to the application of fertilizer [1] and the decline in groundwater level and spring volumes [2] are serious issue. The land use changes caused the decrease in groundwater recharges. In 1965, the total recharge area such as paddy fields, upland fields, grasslands, and forests was 947.89 km^2 [3]. However, that area significantly decreased to 844.30 km^2 in 1991. For example, the area of paddy fields decreased from 262.22 km^2 to 236.56 km^2. The prime reason of decrease in paddy fields is the rice production adjustment policy of Japanese government [2] and urbanization. To improve this situation, a project aiming to increase groundwater recharges started in 2004. In this project, the local famers flood their fields during the fallow periods. Consequently, the groundwater levels and spring volumes have been gradually recovering [2].

In 16 April 2016, the earthquake (main shock) of M_w 7.0 attacked Kumamoto region. Because of the earthquake, the unusual changes in groundwater level were observed. According to Hosono et.al [4], the water level drop was caused by most likely water transfer downwards through open cracks. On the other hand, the coseismic abnormal water level increase was explained by mountain water release [5]. This abnormal water level rise propagated to the downstream water level. It is important to build a model that can express such groundwater level changes and their propagation processes in order to predict water level changes due to future weather conditions and land uses. One of the important matters to build a model is the distribution of groundwater recharge and not-recharge areas classified into each land cover. Thus, in this study, firstly we attempted to generate a land cover map by using temporal Normalized Difference Vegetation Index (NDVI) changes and Geographic Information System (GIS) data. NDVI has been often applied to classification of land cover and assessment of land cover changes [e.g., 6, 7]. In addition to land cover, understanding land use is also an important matter. For example, there are some places in the grasslands where cattle are grazed, while only open burning are performed in the other area. Infiltration tests that we practiced by the double-ring method revealed that the grasslands in which grazing was performed had a lower infiltration capacity. Therefore, based on the results of land cover classification, secondly, we quantified and evaluated the difference of infiltration and surface runoff amounts due to the difference in land uses of grasslands.

2. Materials and Methods

2.1. Study area

Figure 1 shows the analysis target area. The analysis area is located in the center of Kyushu Island, Japan. This area constitutes of 11 municipalities, which has total 1,245 km^2. This analysis area is larger in the north than the area generally called the Kumamoto region. It is because that it can respond to any model area. Annual averaged precipitation and temperature from 1990 to 2019 in Kumamoto observatory were 1,990 mm and 17.2 $^{\circ}$C, respectively [8]. June and July are rainy seasons, often causing heavy rainfall disasters.

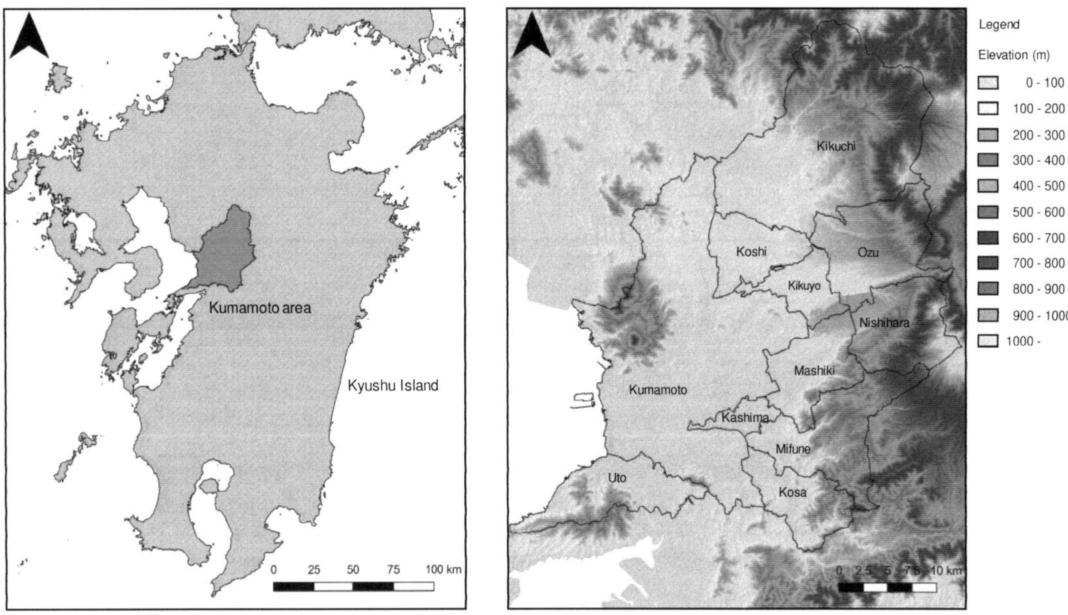

Figure 1. Location map of the Kumamoto area in Japan.

2.2. Satellite images

Three Landsat-8 satellite images on 16 January, 20 March, 23 May 2016 were downloaded from LandsatLook Viewer [9]. We excluded satellite images at the other periods in 2016 because they were covered by clouds. Landsat-8 was launched in 2013 and has two observation sensors: Operational Land Imager (OLI) of 30 m resolution, and Thermal Infrared Sensor (TIRS) of 100 m resolution. The NDVI was calculated by equation (1) using the band 4 (visible red band) and band 5 (near-infrared band) of OLI sensor.

$$\mathrm{NDVI} = (NIR - R) / (NIR + R), \tag{1}$$

where *NIR* is the reflectance of the near-infrared band, and *R* is the reflectance of the visible red band. The NDVI outputs the values between -1.0 and 1.0 which represents the activity level of vegetation.

2.3. Classification of land cover

Kumamoto area (Figure 1) was classified into 9 categories: Grassland, Paddy field, Damaged paddy field, Forest (Conifer and Hardwood), Upland field, Bare land, Building, Golf, Water area (river, lake, and pond). Damaged paddy field means the field which cannot be used because of the 2016 Kumamoto earthquake. The locations of Damaged paddy field were identified by using an account book of paddy field that the address and crop species of each filed were registered and Navigation system [10] which searches the location of field by address. Golf and Water area were identified by using google map. The classification of the other land covers was performed by using temporal NDVI change and GIS. Firstly, the NDVIs were extracted for each land cover by using the polygons of vegetation map (vg67) [11]. Vegetation map (vg67) is a 1/25,000 scale map of land cover, and has been made and updated by field surveys and interpretation of satellite images since 1999. Figure 2 shows extracted NDVIs in mean values for each category. Since Conifer and Hardwood represented almost same trends, it is too difficult to distinguish these land covers. Thus, Conifer and Hardwood were classified as Forest. In the Paddy field, the NDVI decreased from March to May because wheats are harvested before rice planting begins. The NDVI of Grassland cannot show seasonal change like Conifer and Hardwood because of open burning (Noyaki) in March. Utilizing such characteristic changes depending on land cover, the classification was carried out by the following procedure;

1) Grassland; within the grassland area of the vegetation map (vg67), the difference in NDVI between May and March is 0.243 or more.
2) Building; within the building area of the vegetation map (vg67 and vg3) and high-resolution land use and cover map of Japan Aerospace Exploration Agency (JAXA) [12], the NDVI on May is less than 0.319.
3) Paddy filed; within the paddy and upland fields of the vegetation map (vg67 and vg3), the difference in NDVI between May and March is less than -0.017.
4) Upland field; within the paddy and upland fields of the vegetation map (vg67 and vg3), the difference in NDVI between May and March is -0.017 or more.
5) Bare land; excluding the building area of the vegetation map (vg67 and vg3) and high-resolution land use and cover map of JAXA, the NDVI on May is less than 0.319.
6) Forest; the rest area.

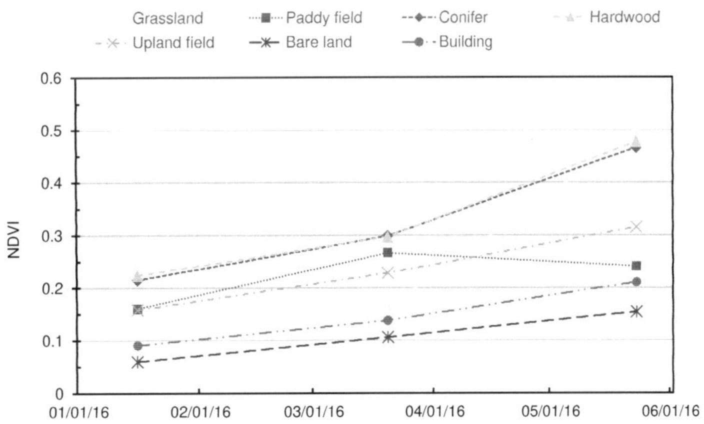

Figure 2. Temporal change of NDVIs for each land cover.

The vegetation map (vg3) is a 1/50,000 scale map of land cover, and was made for 1978-1979, and 1983-1987 land cover [11]. The vg3 was used to complement a small part of the analysis area outside the area of vg67. High-resolution land use and land cover map is the averaged land cover for 2014-2016, which generated by JAXA and has 30 m resolution. According to the Confusion Matrix, the overall accuracy is 81.6 % for 10 categories [12]. In the least accurate category, the accuracy is 56.3%.

2.4. Soil surface model

In this study, we applied the soil surface model (Figure 3) [13], which expresses surface runoff, retention and infiltration, to grasslands area in the outer rim. Infiltration I from the soil surface can be expressed as following equation;

$$I^i = \min\left(P^i + x^i, I_c\right), \tag{2}$$

where P is the precipitation (mm), x is the storage of soil surface model (mm), I_c is the infiltration capacity (mm/hour), and i is the time. Time steps of the model were assigned to 1 hour. When the sum of rainfall and storage exceeds infiltration capacity ($P^i + x^i > I_c$), the water is retained in the soil surface model expressed as following equation;

$$x^{i+1} = \min\left(P^i + x^i - I_c, S\right), \tag{3}$$

where S is the storage capacity (mm). The water exceeding storage capacity is discharged as the surface runoff D. The infiltration capacity was referred to the data [14] in Aso caldera adjacent to the east. Depending on the locations, the infiltration capacity varies from 40 to 118 mm/hour. As mentioned above, the infiltration capacity in the grasslands where grazing has been performed is significantly low. Thus, the calculations were performed in two cases with the high infiltration capacity of 118 mm/hour and the low infiltration capacity of 7.8 mm/hour. According to the soil composition investigated by column sampling of undisturbed soil in the grasslands of Nishihara Village [15], the averaged porosity was 62.9 %. Converted from the column volume and cross-sectional area, the porosity corresponds to 64.9 mm. Generally, since the effective porosity is lower than porosity, the half of porosity was used as the storage capacity. For precipitation, the data at the precipitation observatory where is the closest to the location of classified grasslands was used [16].

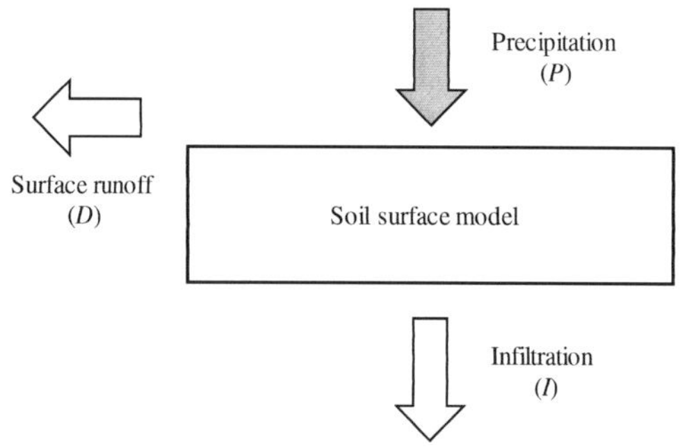

Figure 3. Summary of soil surface model.

3. Results and discussions

3.1. Land cover

Figure 4 shows the results of land cover classifications, and the high resolution land use and cover map of JAXA. The distributions for each category of classification result and JAXA's map are almost the same. However, there is not the category of Golf and Damaged paddy field in JAXA's map. Golf seems to be classified into grasslands. In the center of Ozu and Kikuyo region, actually, paddy agriculture could not be practiced because of the broken irrigation canal induced by the 2016 Kumamoto earthquake. However, our classification method classified these regions into paddy filed. The main reason of this misclassification is that the NDVI change of wheat cultivated before rice was applied because the available satellite image was limited. In paddy fields, the characteristics of NDVI changes are appeared in rice planting and rice growth seasons [17]. Thus, as satellite images in these periods seem to be able to modify paddy fields classification, we plan to purchase SPOT-6/7 satellite images with high resolution of 6 m. The differences in classification results are noticeable around rivers, lakes, and buildings (Figure 5). In the JAXA's map (Figure 5 (b) and (d)), the areas surrounding rivers and lakes are misclassified into the paddy and upland fields. In fact, the riverbed is covered with a vegetation and the area around the lake is a park. Our results successfully classified these areas into forests or bare lands (Figure 5(a) and (c)). Then, JAXA's map could not express the buildings lined up along the fields near the coast (Figure 5(f)). Instead, the paddy fields are distributed. Such misclassification of the paddy fields can overestimate groundwater recharge because the paddy fields are flooded to grow rice, so that water always infiltrates underground regardless of precipitation. In order to evaluate a validity of our classification results, the area of each category was compared with JAXA's map (Figure 6). In addition, the area of Golf was added to the area of Grassland, and the area of Damaged paddy field was added to the area of Paddy field. As mentioned above, although the misclassification of JAXA's map were confirmed, the overall accuracy of classification is high. Although the coefficient of determination R^2 of 0.94 was obtained, there is the significant difference in Paddy field and Building. The area of paddy fields of JAXA's map is higher than that of our classification mainly because of the misclassification as shown in Figure 6. On the other hand, the area of building of our classification is higher than that of JAXA's map. The difference of distribution can be observed especially in the paddy fields of south-west area and north area (Figure 4). It is because that JAXA's map could not extract slender and/or small buildings area between paddy fields (Figure 5(e) and (f)).

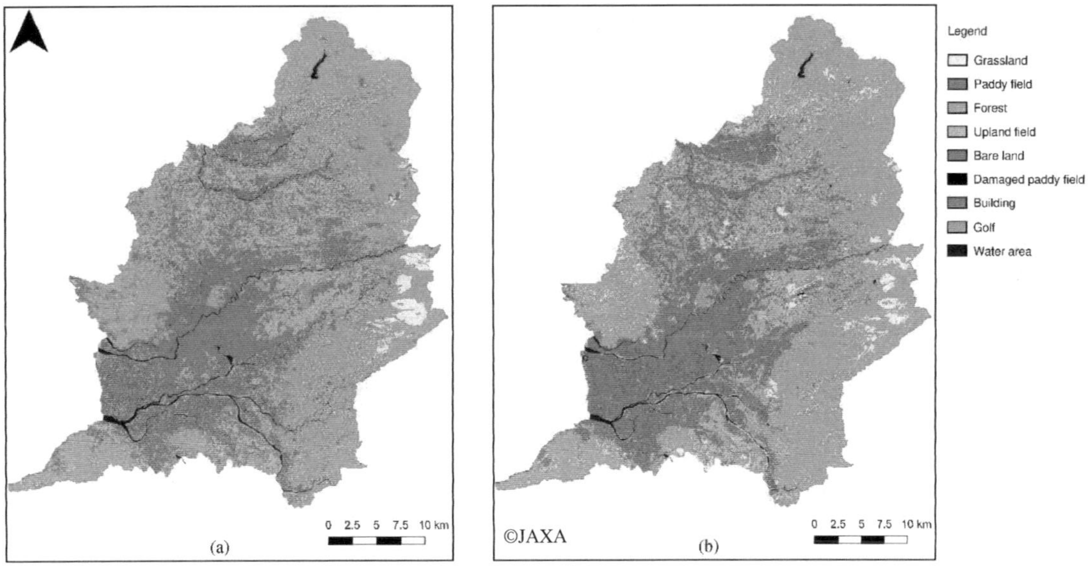

Figure 4. Land cover map: (a)NDVI and GIS, (b)JAXA*.

*Provided by high-resolution land use and land cover 30m resolution map of Japan [2014 – 2016] (ver.18.03: 10 categories) (Japan Aerospace Exploration Agency)

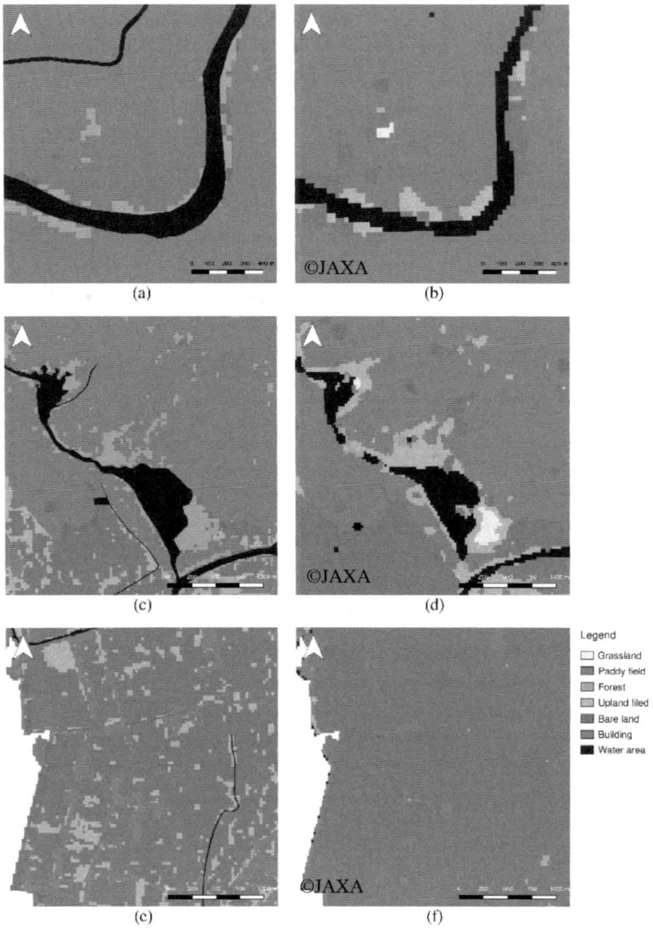

Figure 5. Comparison of land cover map: (a)NDVI and GIS, (b)JAXA, (c)NDVI and GIS, (d) JAXA, (e)NDVI and GIS, (f)JAXA

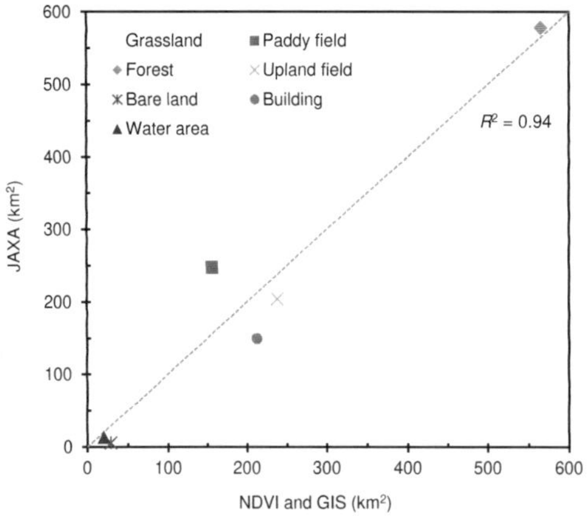

Figure 6. Relationships between classification results and JAXA's map.

3.2. Infiltration and surface runoff

Figure 7 shows the calculated infiltration and surface runoff amounts in two patterns from 1 June to 31 July (rainy season in Japan). As shown in Figure 7(a), when the infiltration capacity was assigned to

118 mm/hour, the runoff did not occur, which means that all of precipitation infiltrates into underground. On the other hand, when the infiltration capacity was assigned to 7.8 mm/hour, the runoff occurred in some precipitation events. The cumulative surface runoff amount reached about 2.1 million m^3. These calculations show the difference in land use of grasslands as an example, but the difference in land use is also recognized in the building area. The building area is usually considered as an impermeable area. However, a part of site in Kumamoto prefectural office, a permeable paving is applied to promote infiltration of precipitation. Namely, if the land cover is the same but the land use is different, it is important to assign the parameter values corresponding to the land use when a model is created.

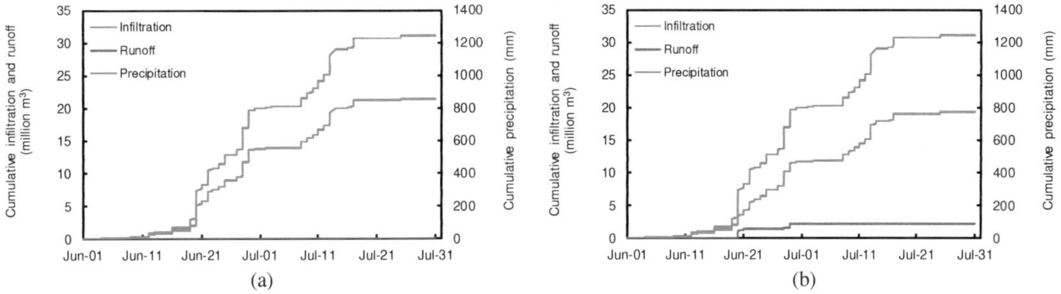

Figure 7. Cumulative infiltration, runoff, and precipitation from 1 June to 31 July 2016: (a)Infiltration capacity is 118 mm/hour, (b)Infiltration capacity is 7.8 mm/hour.

4. Conclusions

In this study, we classified Kumamoto area where groundwater is the important water resource into 9 land cover categories for 2016 by using temporal NDVI change depending on each land cover and GIS data. The classification results are almost corresponding to the land cover map of JAXA. In some places, our classification results well represented the actual land covers. However, the paddy fields were misclassified because of the limited satellite images so that we plan to purchase SPOT-6/7 satellite images with high resolution of 6 m in the suitable periods to classify paddy fields. The soil surface model indicated that the different land use should be considered in the same land cover to evaluate infiltration and surface runoff amount.

5. References

[1] Hosono T, Tokunaga T, Kagabu M, Nakata H, Orishikida T, Lin I-T and Shimada J 2013 The use of $\delta^{15}N$ and $\delta^{18}O$ tracers with an understanding of groundwater flow dynamics for evaluating the origins and attenuation mechanisms of nitrate pollution *Water Res.* **47(8)** 2661-2675.

[2] Ichikawa T, Nakagawa K and Berndtsson R 2020 Earthquake effects on artificial groundwater recharge efforts in south Japan *Environ. Earth Sci.* **79** 142.

[3] Kumamoto Prefecture and Kumamoto city 1995 Report on groundwater comprehensive survey in Kumamoto region, 122p.

[4] Hosono T, Yamada C, Shibata T, Tawara Y, Wang C-Y, Manga M, Rahman A T M S and Shimada J 2019 Coseismic groundwater drawdown along crustal ruptures during the 2016 M_w 7.0 Kumamoto earthquake *Water Resource Res.* **55(7)** 5891-5903.

[5] Kagabu M, Ide K, Hosono T, Nakagawa K and Shimada J 2020 Describing coseismic groundwater level rise using tank model in volcanic aquifers, Kumamoto, southern Japan *J. Hydrol.* **582** 124464.

[6] John R, Chen J, Lu N and Wilske B 2009 Land cover/land use change in semi-arid Inner Mongolia: 1992-2004 *Environ. Res. Lett.* **4** 045010.

[7] Lunetta R S, Knight J F, Ediriwickrema J, Lyon J and Dorsey Worthy L 2006 Land-cover change change detection using multi-temporal MODIS NDVI data *Remote Sens. Environ.* **105** 142-154.
[8] Japan Meteorological Agency. Weather observation data. Available online: http://www.data.jma.go.jp/obd/stats/etrn/index.php?prec_no=86&block_no=47819&year=&month=&day=&view= (accessed on 10 August 2020).
[9] U.S. Geological Survey. LandsatLook Viewer. Available online: https://landsatlook.usgs.gov/ (accessed on 17 December 2019).
[10] Agriculture land information system agriculture committee. Available online: https://www.alis-ac.jp/ (accessed on 8 July 2020).
[11] Biodiversity center of Japan. 1/25,000 vegetation survey (6th). Available online: http://gis.biodic.go.jp/webgis/sc-006.html (accessed on 31 January 2020).
[12] Japan Aerospace Exploration Agency. High-Resolution Land Use and Land Cover Map of Japan (Version 18.03). Available online: https://www.eorc.jaxa.jp/ALOS/en/lulc/lulc_index.htm (accessed on 28 July 2020).
[13] Asia Air Survey CO., LTD. 2013 Report on investigation of the impact on river flow due to forestation of grasslands in Aso area in 2012, 102p.
[14] Takano Y, Uemura T, Murakami M and Yoshimura K 2015 Long term change in flow rate of spring at Otomeyama Park in Shinjuku Ward and its possible cause *J. Groundwater Hydrol.* **57(2)** 171-185.
[15] Hamada S 2001 Basic study on compositional change of surface soil during formation process of groundwater recharge forest *Graduate school of Kyushu Tokai University Master thesis*, 65p.
[16] Ministry of Land, Infrastructure, Transport and Tourism. Water information System. Available online: http://163.49.30.82/cgi-bin/SiteInfo.exe?ID=109101289908040 (accessed on 4 August 2020).
[17] Amano H, Iwasaki Y and Ichikawa T 2020 Identification of paddy fields using temporal NDVI changes in Minamiaso Village, Kumamoto Prefecture, Japan to estimate groundwater recharge. *IOP Conf. Ser.: Earth Environ. Sci.* (in press).

Acknowledgments
This research was performed by the Environment Research and Technology Development Fund (JPMEERF19S20505) of the Environmental Restoration and Conservation Agency of Japan.

Water Quality Improvement Effective Analysis for Artificial Groundwater Recharge in Urban River Network System

Wang Xiao[1], Wei Jun[2]*,Cheng Kaiyu[3], Zhao Mingzhi[4]

1 Lecturer, Ningbo University, China
2 Professor level senior engineer, Huadong engineering corporation limited, China
3 Professor level senior engineer, Huadong engineering corporation limited, China
4 Senior engineer, Huadong engineering corporation limited, China
E-mail: wangxiao1@nbu.com

Abstract. With increase of population and urbanization, the persistent deterioration of urban rivers has become the most intractable problem for environment management agency. After the boom of pollution control projects implementing such as sewage system and sewage treatment plants building to decrease the pollution load, there is a growing interest on increasing ground water discharge with artificial groundwater recharge work and improving load capacity of target river. Under this background, it is important to study the exact effect of water artificial groundwater recharge work on water quality improvement, especially in the river network area, to give support on exact designment. In this study, the river network in Hanjiang district featured with complex river network has been selected as study area. Based on water quality monitoring result and current load statistical result, a one-dimensional mathematic model of target river network has been set up and the water quality has been simulated with different water artificial groundwater recharge schemes, which considering both river supplementary discharge and water quality factors. Then the simulation results under different effects have been analysed and the optimal scheme has been selected to meet the goals. The analysis results indicate that qualified option is recharging 2 m^3/s fresh water of level III, under which, the water quality of study area can achieve the basic standard of Level V as qualified scenic water.

1. Introduction

According to the investigation of the current situation, the main problems affecting the water quality improvement of urban river courses are as follows: (1) both economy development and population growth in urban area have bring in large amount of urban residential pollution, industry pollutions and runoff pollution, which means more pollution will discharged into rivers than before. (2) most of urban streams are in small scale, both the river discharge and the load capacity of urban rivers is really limited. That means only small amount of pollutants will bring in the water deterioration in urban rivers [1-3]. To deal with these two key problems, both pollution control projects and ecological water artificial groundwater recharge projects should be comprehensively implemented to help urban river water environment restoration.

In recent years, many cities have made great efforts on water quality improvement and river water artificial groundwater recharge work has become one of the most important measures. The Cheonggyecheon Restoration work [4], an ambitious initiative to restore a 5.8-kilometer stream in central Seoul, is the most famous and typical unban river restoration work in the world. Artificial groundwater recharge projects, as one of massive civil engineering projects in Cheonggyecheon Restoration Project, was designed and built to deliver 120,000m^3/s clean water from other basin to

IOP Conf. Series: Earth and Environmental Science **728** (2021) 012003 doi:10.1088/1755-1315/728/1/012003

Cheonggyecheon stream, which help to develop the water quality and launch the broader regeneration of Seoul's inner areas.

According to the experience of water artificial groundwater recharge projects, the main benefits of river artificial groundwater recharge are as follows: (1) the water artificial groundwater recharge work will enhance the load capacity of rivers dramatically and develop the load degradation ability, which could improve the water quality of urban rivers; (2) it will maintain the ecological baseflow of unban streams and help rehabilitate rivers ecosystem. Attribute to these benefits, the recharge project has been introduced into more and more unban rivers restoration projects. Since 1982, paul has studied the organic water quality changes during groundwater recharge in palo alto baylands[5]. In 2007, Reija E. Kolehmainen has studied and evaluated the removal of natural organic matter (NOM) and structural changes in the rnicrobial community during infiltration of humic lake water at three artificial groundwater recharge (AGR) sites in Finland[6]. Under this background, it is important to study the exact effect of water artificial groundwater recharge work on water quality improvement, to give support on exact artificial recharge system designment and management.

2. Background and basic data

2.1. Study area

Hanjiang District, located in the eastern coast of Putian City, Fujian Province, is one of the central urban area with rapid development of industry and economy. The total area is 752 km^2 and the population is over 440 thousand. Hanjiang district is featured by high density of river network. The main rivers include Xikou River, Wuzi River, Wangjiang River, Tangtou River, and Haicen River. Among them, Xikou River, Wuzi River and Tangtou River are the main drainage channels, and Haicen River is the important drainage channel for Dongzhen Reservoir, an important water storage project in Putian City. All rivers converge at low-lying areas downstream and form crisscross river network. The river network in Hanjiang district is shown in Figure1.

Attribute to the high speed of urbanization and industrial development in downtown area, large number of sewage and pollutants go into the rivers directly and the amount has exceeded the capacity dramatically. According to the monitoring results of water quality in 2017, the waters in Tangtou River, Wangjiang River, Wuzi River and Gongkou River in study area should all be graded into moderate black and odorous water according to the grading standard, and the most obvious indicators are ammonia nitrogen and total phosphorus. That means it is far from the water quality standard for water environment management in Hanjing district.

Since 2018, with sewage system of whole Hanjiang district developing, the pollution load has been dramatically controlled. However, due to the limited load capacity, lots of urban rivers still failed to reach the expected standard of grade V, which is the lowest level for urban scenic water. Then the water artificial groundwater recharge project was proposed to expand the load capacity and help water environment restoration.

In order to study the exact effect of water artificial groundwater recharge on water quality improvement in river network and identify the most reasonable recharge scheme in Hanjiang central district, the water quality under different artificial groundwater recharge conditions will be simulated by using the constructed regional river network model.

2.2. Pollution load statistics

The pollution in Hanjiang central area can be mainly divided into point source pollution and non-point source pollution according to the different way of sewage discharge. The point source pollution mainly includes the residential pollution and industrial pollution discharging directly in the scope of the study area. The non-point source pollution is mainly including the urban runoff pollution and a small amount of scattered living and farming pollution.

According to the investigation and statistical results, the ammonia and TP emissions in the study area in 2017 were 231.9t/a and 32.92t/a respectively. Details of result are shown in Table 1, and the proportion of various pollution sources is shown in Figure 2.

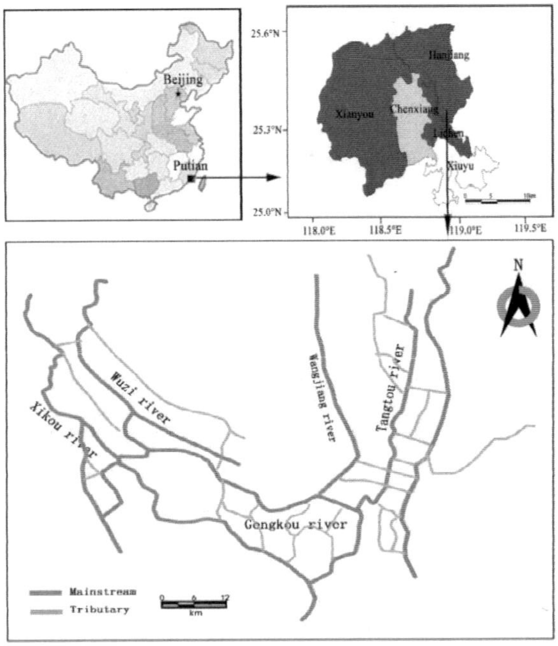

Figure 1. Location and overview of river network in Hangjiang central area

Table 1. Pollution load statistical results **unit: t/a**

pollutants	POLLUTION LOAD				
	industry	urban residential	livestock	farming	runoff pollutions
NH_3-N	43.93	166.25	5.51	4.92	11.34
TP	10.21	19.5	0	0.98	2.27

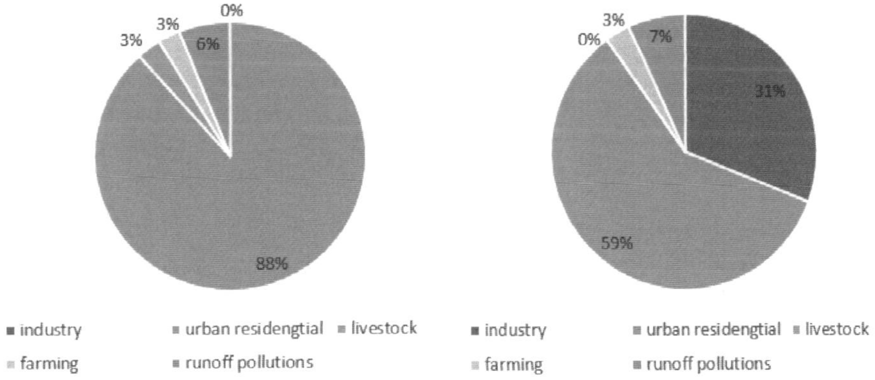

（a）Ammonia-nitrogen（NH_3-N） （b）Total Phosphorus (TP)

Figure 2. Pollution load proportion of different kinds of pollutants

It can be investigated that the domestic pollution source contributes the most to the regional pollution load, accounting for 57-70% of the total load, followed by the industrial direct pollution source, accounting for 21-33% of the total. The pollution load from the nonpoint sources, including living, farming and urban runoff sources contribute the least, accounting for only 1-7% respectively.

Therefore, it can be concluded from the analysis that the pollution control targeting on the residential and farming pollution sources should be pay more attention to ensure the improvement of water quality in study region.

According to the water protection planning in Hanjiang district, the sewage system will be developed to collect more industry pollution, urban residential pollution and runoff pollution for pollution control at the beginning of pollutants migration. Cooperated with planning, the river dredging work and eco-restoration projects have also been planned to control pollution load at the end step of pollutants migration. According to the planned scale of different projects, the amount of pollution control estimated was shown as below in Table 2. Combined the current pollution load with estimated pollution control amount, the exact pollutants discharged in the river was 49.25 NH_3, 8.76 TP.

Table 2. Estimated amount of pollution control **unit: t/a**

pollutants	**POLLUTION CONTROL AMOUNT**		
	sewage system	restoration projects	river dredging work
NH_3-N	162	12	8.7
TP	23	-	1.2

2.3. Pollution load statistics

Based on the investigation and field survey of the river system, the mathematical model could be built to quantify the effects of water quality improvement. Considering the hydrodynamic and pollutants transportation characteristics of the river network in study area, one-dimensional river network modelling was built using MIKE11 software. Saint-Venant Equation was the calculation formula for hydrodynamic calculation as followed:

$$\frac{\partial A}{\partial t}+\frac{\partial Q}{\partial x}=q \qquad \text{(Formula 2.3-1)}$$

$$\frac{\partial Q}{\partial t}+\frac{\partial(\alpha\frac{Q^2}{A})}{\partial x}g+gA\frac{\partial h}{\partial x}+\frac{gn^2Q|Q|}{AR^{\frac{2}{3}}}=q \qquad \text{(Formula 2.3-2)}$$

where A, R denote the cross-section area and hydraulic radius of target cross section; x and t denote the distance in downstream direction (m) and time (s); Q and q are the discharge (m^3/s) of main stream and lateral inflow (m^2/s); g is the gravity acceleration; α is the momentum correction coefficient.

On that base, the advection and dispersion of pollutants along rivers can be simulated combing the hydrodynamic simulation results using the formula as followed:

$$\frac{\partial C}{\partial t}+\mu\frac{\partial C}{\partial x}=\frac{\partial}{\partial x}(E_x\frac{\partial C}{\partial x})-Kc \qquad \text{(Formula 2.3-3)}$$

Where c denotes the pollution concentration； u denotes the average velocity； E_x is convection diffusion coefficient; K is comprehensive degradation coefficient;

This modelled system includes 24 river segments, covering all important rivers in study area as shown in Figure 3. the river section spacing varies from 200 to 1000 meters and he total number of computational grids is about 196.

The basic parameters of the model were calibrated using the water quality monitoring data measured in 2016. It is determined that the roughness coefficient of the river bed is 0.025-0.038 and Manning number is 28-40. Besides, the water quality parameters of degradation coefficient are 0.06-0.13 D-1 for ammonia nitrogen-N, 0.03-0.09 D-1 for total phosphorus. It can be seen from the comparison results of water quality calculation results and measured values that both are in good agreement. Therefore, the river network model of this project can accurately simulate the water quality situation of Hanjiang project area.

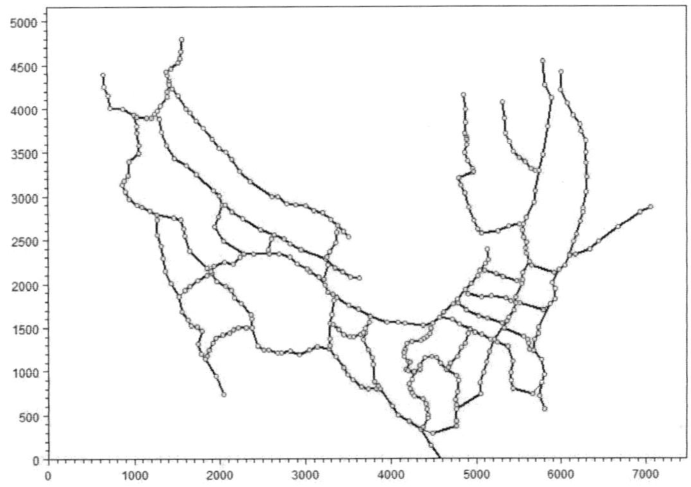

Figure 3. River network of mathematical model in study area

3. Simulation and results

The "water function zoning plan of Putian city" required the water quality of the urban rivers in study area achieving grade V. According to water quality analysis result, the key factors are NH_3-N and TP. Therefore, ammonia nitrogen and total phosphorus were considered as the pollution control indicators in this study.

In order to study the exact effects of water artificial groundwater recharge work on water quality improvement and identify the most reasonable recharge scheme, the water quality under different recharge conditions will be simulated by using the constructed regional river network model. Based on the water resource survey in the Hanjiang district, the recharge path and scale could be designed. The cleaning water could be recharged from the upstream of Diluxi and the maximum available recharge flow is 4 m^3/s. So different boundary conditions for prediction could be seen in Table 3 for details. And the simulation results of different scheme could be detailly shown in below.

Table 3. Details of designed water artificial groundwater recharge schemes

Schemes	Discharge (m^3/s)	Water quality	Contaminate concentration (mg/L)	
		-	NH_3-N	TP
A	0	-	-	-
B	2	Level IV	1.5	0.3
C	4	Level IV	1.5	0.3
D	4	Level III	1	0.2

Scheme A: without any clean water recharged, the water quality of main streams is mostly in the grade V- and east part of river network has been polluted more seriously; the key pollutant indicator is ammonia nitrogen and concentration for average ammonia nitrogen in east part is as high as 5.2mg/l, which is tribble times larger than the standard.

Scheme B: after implementation of water artificial groundwater recharge work for the scale of $2m^3/s$ with water quality in Level IV, the water quality of whole river network is significantly improved. The concentration for ammonia nitrogen in east part is 4.2mg/l and for TP is 0.52mg/l.

Scheme C: after improving the recharge scale to $4m^3/s$, the water quality of whole river network is significantly improved. The average concentration in east part for ammonia nitrogen is 3.2mg/l and for TP is 0.48mg/l. The pollutants concentration for TP could reach the requirements standard.

Scheme D: after strengthening the water artificial groundwater recharge work via improving the water quality of recharged water into level III , the concentration of main streams could be improved in the range of 1.5- 2.2mg/l for ammonia nitrogen and 0.41 for TP. Both the concentration of ammonia nitrogen and TP could reach the requirements of class V standard. That means the rivers can basically reach the scenic water under that water artificial groundwater recharge scheme.

Based on the caparison results, the scheme D is the only option which could meet the water quality requirements. That means both the recharge work and the water quality maintaining work of upstream should be implemented.

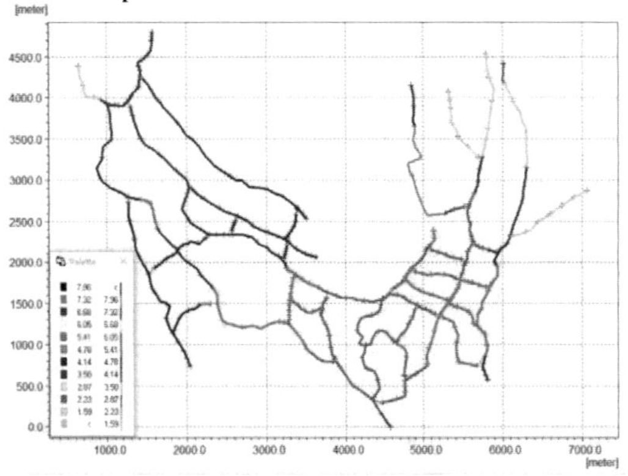

(a). SCHEME A

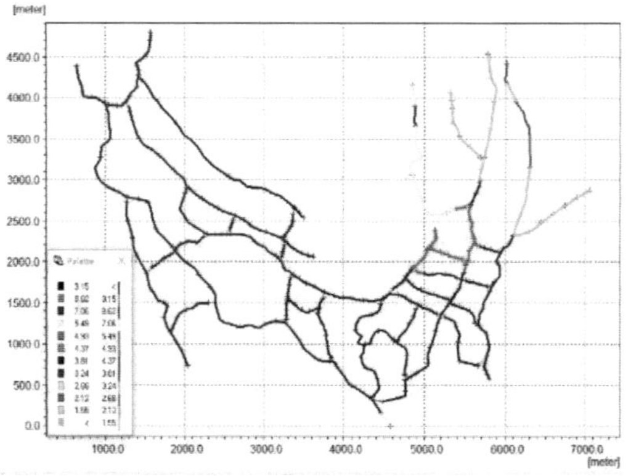

(b). SCHEME B

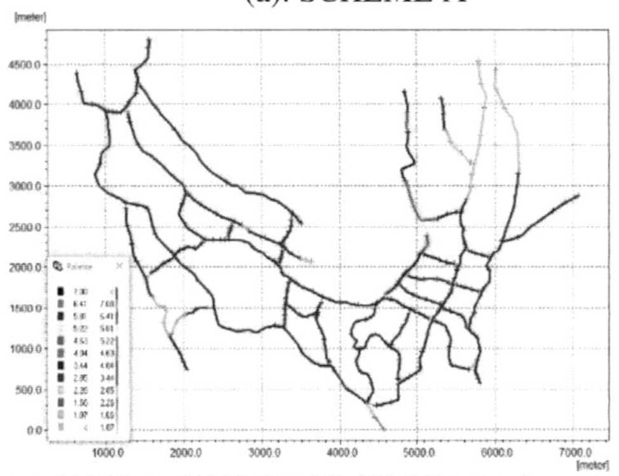

(c). SCHEME C

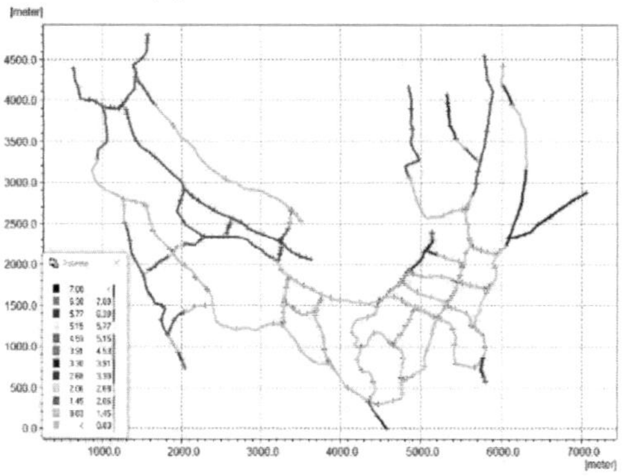

(d). SCHEME D

Figure 4. Water quality simulation results of river network for NH_3

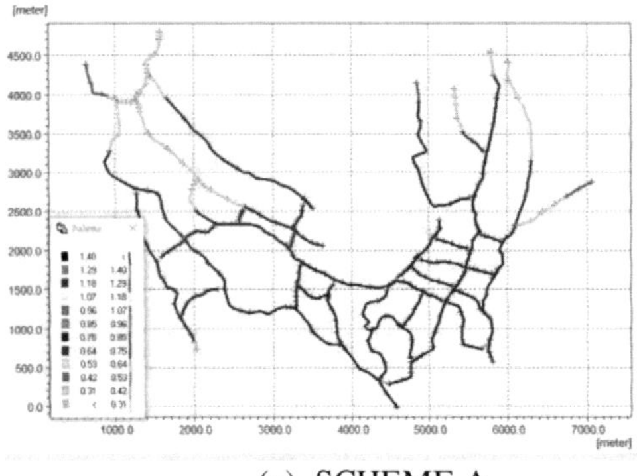

(a). SCHEME A

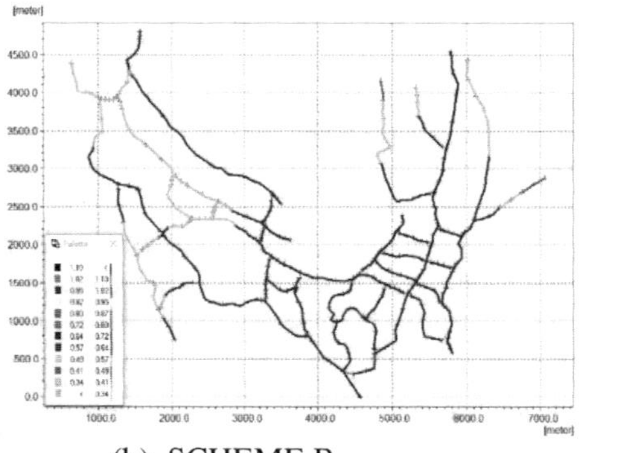

(b). SCHEME B

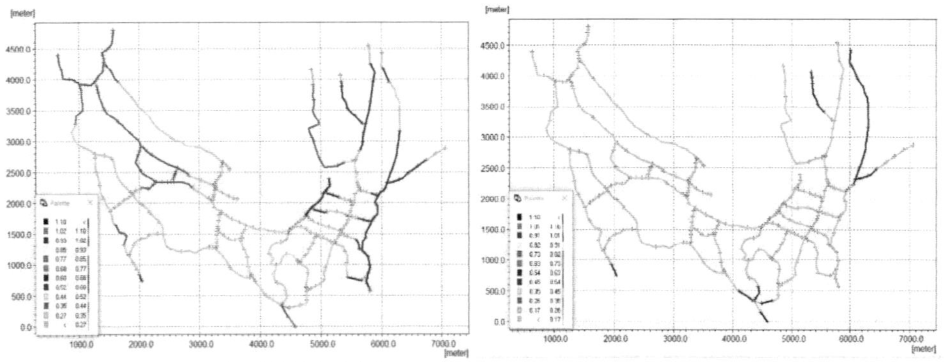

(c). SCHEME C (d). SCHEME D

Figure 4. Water quality simulation results of river network for TP

4. Discussion

According to the simulation results of different schemes, the water quality of whole river network has been improved dramatically attributing to the water recharge work, especially in the east part, the most polluted area. Taking ammonia nitrogen as key pollution factor, the detailed water quality variation trend for main streams in river network could be analysed as below:

Wangjiang river, as the most polluted river, is chosen to implemented recharge work directly for pumping in 1.5-3 m^3/s fresh water at the upstream. As Figure 5-(a) shown, in scheme A, the concentration at cross section of 500m length is as large as 7.35mg/L, then it continuously decreases to 5.93 mg/L as the blue line dropping. That means before the recharge work, the water quality of wangjiang river is affected heavily by the seriously polluted inflow than the locally discharged pollutants. In scheme B, C and D, the heads of concentration lines have dropped in different degrees attributing to water recharge works at the upstream of Wangjiang river. Then the concentration lines go in different trends as the length changing. Especially in the scheme D, with the recharge work of the highest level, the concentration of inflow has archived the requirements of 2mg/L, then the concentration line increases slightly, which means under this scheme the polluted inflow made less contribution on water quality of Wangjiang river and more pollution control projects should be implemented to control the local load and fit the load capacity of Wangjiang river.

In Figure 5-(b), the heads of all the lines is in the range of 2.12-2.85mg/L, that means for Tangtou river, the inflow water quality is in a relevant satisfied state. As the length growing, there is an obvious rise at the cross section of 2500m of Tangtou river. That was because the Wangjiang river and tangtou river flow converge here, the water quality of Tangtou river has been influenced by the polluted water from Wangjiang river. However, from the comparison of different scheme lines of A, B, C and D, it could be clearly seen that, as the water quality of Wangjiang river continually improving, the concentration of whole Tangtou river will reach the requirement. That means before the recharge work, the water quality of Tangtou river was affected more by the seriously polluted inflow of Wangjiang than the locally discharged pollutants.

Wuzi river was another stream recharged clean water from the upstream. Similar to Wangjiang river, the heads of concentration lines in figure5-(c) have dropped dramatically attribute to different water recharge works. And with the cleanest water recharged, the concentration of inflow has archived the requirements of 2mg/L, also the slightly rising concentration line illustrated more pollution load should be controlled to fit the load capacity.

Gongkou river was the stream in the downstream connecting the other main rivers, that means the water quality has been affected by the Tangtou river and Wuzi river directly. As the figure 5-(d) shown, the variation amount of concentration lines was in a quite small range and the concentration of NH_3 has been in scale of level V steadily.

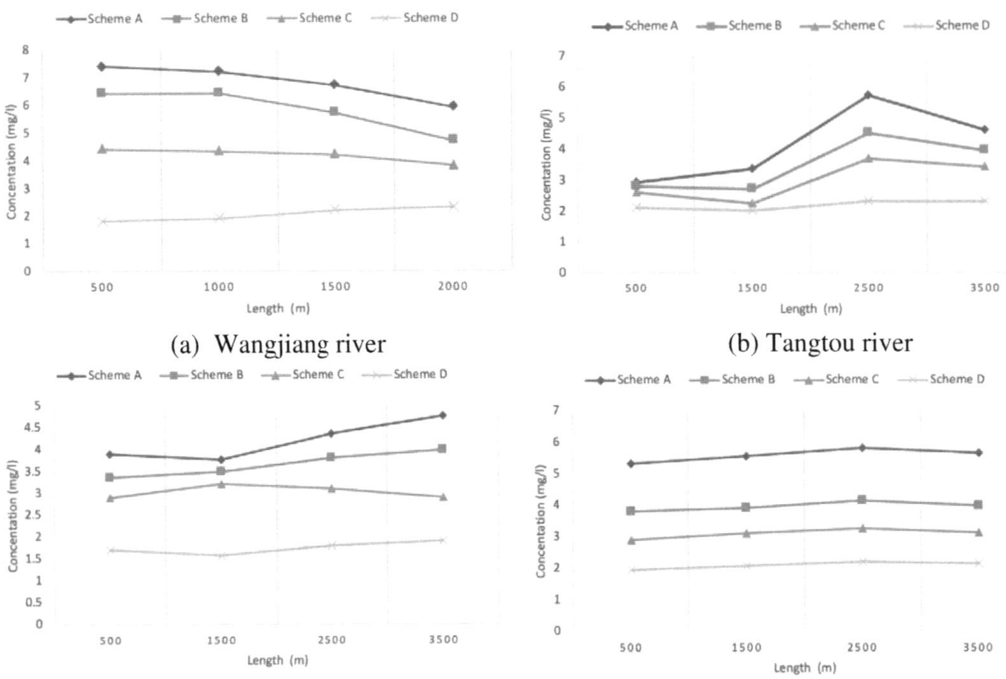

(a) Wangjiang river (b) Tangtou river

(c) Wuzi river (d) Goukou river

Figure 5. Water quality simulation results of river network for NH_3

5. Conclusion

In this study, a quantified method based on mathematical simulation has been proposed to research the exact effect of river artificial groundwater recharge on water quality improvement. Appling the 1-D mathematic model of river network in Hanjiang district, the water quality of whole river network has been simulated considering different artificial groundwater recharge conditions. Relevantly, the different effects on water quality improvement have been analysed and the best scheme which can help the water quality achieving standard has been chosen. Under that background, the effective artificial groundwater recharge system and relevant projects can be then designed to make contribution for the whole water restoration projects.

The goal of ecological water artificial groundwater recharge work in Hanjiang district is to improve the water quality and help recovering the healthy status of whole water environment. After the water artificial groundwater recharge system being built, the load capacity of urban rivers will be expanded to achieve the effect of improving water quality effectively.

6. References

[1] Zhang Y, Zhang Y, Gao Y, et al. Water pollution control technology and strategy for river–lake systems: a case study in Gehu Lake and Taige Canal[J]. Ecotoxicology, 2011, 20(5): p.1154-1159.

[2] Anonymous. WATER POLLUTION RESEARCH LABORATORY IMPROVES POLLUTION CONTROL TECHNOLOGY[J]. chemistry & industry, 1969(43):1547.

[3] Huang K,Jin Z,Li J,et al.TECHNOLOGIES AND ENGINEERING PRACTICES OF WATER POLLUTION CONTROL FOR A TYPICAL URBAN RIVER IN DIANCHI WATERSHED[J]. Fresenius Environmental Bulletin, 2014, 23(12c):3468-3474.

[4] Choe L J, Jung S W, Kim D G, et al. Temporal changes in benthic macroinvertebrates and their interactions with fish predators after restoration in the Cheonggyecheon, a downtown stream in Seoul, Korea[J]. Entomological Research, 2014, 44(6):338-348.

[5] Roberts P V, Schreiner J, Hopkins G D. Field study of organic water quality changes during groundwater recharge in the Palo Alto Baylands[J]. water research, 1982, 16(6):1025-1035.

[6] Kolehmainen R E, Langwaldt J H, Puhakka J A. Natural organic matter (NOM) removal and structural changes in the bacterial community during artificial groundwater recharge with humic lake water[J]. water research, 2007, 41(12):0-2725.

Acknowledgments
This paper is financially supported by the China Postdoctoral Foundation (208930 Study on Pollution Transfer Simulation and Optimal Control Considering the Characteristics of Urban Complex Water Environment)

IOP Conf. Series: Earth and Environmental Science **728** (2021) 012004 doi:10.1088/1755-1315/728/1/012004

Implementation of an Environmentally Friendly Approach of a Hospital–Practice of Medical Waste Disposal

Chiwei Chen[1]

1 Professor, College of Economics and Management, Zhaoqing University
Zhaoqing, Guangdong 526061, China
E-mail: 3418221808@qq.com

Abstract. Gama Hospital is a well-known private hospital in Taiwan. In addition to medical treatment technology, hospital management and cost control are also excellent. Regarding the treatment of medical waste, one must consider the cost and also the environmental implications. Waste disposal, besides the inevitable treatment after its emergence, should be given more attention during the course of medical execution. In the practice of the hospital, the actual waste treatment cost was reduced. The main factor is that the hospital promotes a series of waste treatment education and training. In addition to medical personnel, patients and family members also participate in medical waste reduction and correct handling activities, successfully completing the hospital's waste reduction target and achieving sustainable environmental social responsibility.

1. Introduction

The World Health Organization (WHO) has pointed out that about 85 per cent of the total waste generated in medical activities is generally harmless waste, while the remaining 15 per cent is considered to be a hazardous substance of pollution, chemical or radioactivity [1]. Globally, 16 billion injections are made worldwide, but not all needles and syringes are properly treated after use, and in some cases, open burning of medical waste can lead to toxic air pollution emissions. The aim is to ensure that safe and environmentally harmless management measures for medical waste are in place to prevent adverse health and environmental effects.

Common problems with the disposal of medical waste are as follows: lack of awareness of health hazards associated with medical waste, inadequate training on waste disposal, lack of waste management and disposal systems, and low attention to this issue, as well as inadequate financial and human resources. In low and middle-income countries, there has been a significant reduction in the use of contaminated needles and syringes in recent years, partly due to efforts to reduce the reuse of syringes. In 2010, however, unsafe injections resulted in as many as 33,800 new HIV (Human Immunodeficiency Virus) infections, 1.7 million hepatitis B infections and 315,000 hepatitis C infections [2].

According to WHO statistics [3], each hospital bed in industrialized countries produces 0.5 kg of medical waste per day per bed, while in developing countries each bed produces 0.5-2.5 kg of medical waste per day. The United States of America produces about 2.6 million tons of medical waste each year. Therefore, each hospital should be particularly and continuously attentive about the generation and treatment of medical waste. The methods and data presented in this paper are the result of Gama Hospital's case. Gama Hospital aims to make a minor contribution toward environmental friendliness and environmental sustainability.

2.Gama Hospital promoted medical waste treatment methods

Gama Hospital is a private hospital in Taiwan. In addition to its well-known medical technology, its hospital management and cost control are also excellent. Based on the high priority attached to the treatment of medical waste, Gama Hospital actively promoted the reduction of medical waste. The hospital set short-term and long-term goals. The short-term goal was to reach a specific reduction target, to be socially responsible for the sustainable environment, and to achieve environmental friendliness.

2.1. Classification of medical waste

Medical waste may be largely classified as general medical waste and hazardous medical waste which includes infectious, chemical, and radioactive waste. The WHO also listed common hazardous medical waste as follows [1]:
(1) Infectious waste: Waste contaminated by blood or other liquids.
(2) Pathological waste: Human tissues, organ fluids.
(3) Sharp ware: Needles, syringes and disposable scalpels.
(4) Chemicals: Solvents, disinfectants for laboratory preparation.
(5) Drugs: Expired, unused, and contaminated drugs or vaccines.
(6) Genetically toxic: High-risk, mutation-causing, teratogenic and carcinogenic.
(7) Radioactive waste: Radioactive nuclide-contaminated items.

2.2. The importance and purpose of infectious waste classification

The correctness of classification can affect the treatment process and cost, and due to the characteristics of the infectious waste, classification does not directly affect the safety of hospital workers, but may generate secondary environmental pollution. The purposes of improving treatment of infectious waste are as follows: to strengthen people's understanding of the correct classification of infectious waste, to achieve the effect of reducing the amount of infectious waste, and to reduce the cost of treatment of infectious waste.
The Bureau of Health of Taiwan, the Bureau of Environmental Protection of Taiwan, and other health administration offices of Taiwan have set clear standards for the determination and treatment of medical waste [4-8], and have requested all medical institutions to implement all related regulations in a firm and serious manner, as follows: waste classification, waste characteristics analysis, storage methods, adherence to labels, typical waste disposal, and identification of major sources.
Since the implementation of health insurance in Taiwan, the service amount of various medical institutions has increased and so has the amount of medical waste. The medical waste disposal costs have increased and the operating costs of hospitals have also increased. Gama Hospital will review the current treatment of medical waste and carry out feasible solutions to analyze its driving effect.

2.3. Analysis of the status of branches

The review plans of Gama Hospital include hospitals such as the Linkou, Kaohsiung, Keelung, Jiayi and Yunjianan Regional Hospitals, and will begin with the Jiayi Branch. From Table 1, it can be seen that the daily per bed in the Linkou Branch produces 0.45 kg of infectious waste, which is the smallest number among the branches, while the Jiayi Branch is the largest daily producer of infectious waste. In terms of general waste, the Linkou Branch produces the most at 2.83 kg per bed per day and the Keelung Branch produces the least at 1.87 kg per bed per day. In year A and B, the total amount of infectious waste in the Jiayi Branch was 194 tons and 223 tons respectively, compared with 594 tons and 566 tons for general waste.

Table 1. Medical waste production in each branch

Branch	Infectious waste (kg/bed-day)	General waste (kg/bed-day)
Linkou Branch	0.45	2.83
Kaohsiung Branch	0.49	2.18
Keelung Branch	0.52	1.87
Yunjianan Regional Hospital	0.52	--

Jiayi Branch (year A)	0.68 Total amount (194,659kg/year)	2.06 Total amount (593,694kg/year)
Jiayi Branch (year B)[a]	Total amount (223,060kg/year)	Total amount (566,334kg/year)

[a]Note: A+1=B

In year A the infectious waste disposal cost at the Jiayi Branch was $502,869 (in this paper, currency unit used is NT$, New Taiwan Dollars; exchange rate of NT$ to US$: 30NT$=1US$) per month, and in year B was $537,372 per month (see Table 2), an increase of $34,503 per month. In general waste disposal, the monthly treatment cost in year A was $114,781, while the monthly cost in year B was $116,256 per month, an increase of $1,475 per month. From year A to B, the monthly processing cost increased $35,978.

Table 2. Medical waste disposal costs in Jiayi Branch (Currency unit: NT$)

Year Waste types	Year A (A)	Year B[a] (B)	Cost difference (B) – (A) = C
Infectious waste	194,659 ∗ $31/kg= $6,034,429 $6,034,429/12= $502,869/month	223,060 ∗ $26.5/kg= $5,911,090 $5,911,090/11= $537,372/month	**$34,503/month**
General waste	593,694 ∗ $2.32/kg= $1,377,370 $1,377,370/12= $114,781/ month	566,334 ∗ $2.26/kg= $1,278,815 $1,278,815/11= $116,256/ month	**$1,475/month**
Total amount	$617,650/ month	$653,628/ month	**$35,978/ month**

[a]Note: year B only collect 11 months of data

2.4. Analysis of the current problems of medical waste disposal

In order to deeply explore the source of problems arising from the cost of medical waste disposal in the Jiayi Branch, this paper uses a Cause and Effect diagram (Fishbone diagram) shown as Figure 1. In the diagram, both staff members and the families of patients are under-aware of the classification of medical waste, and staff treatment of infectious waste is not comprehensive. On the environmental aspect, the markings of medical waste sites are unclear. On the policy aspect, there is no audit of the misclassification of infectious waste, and no reward and punishment policy had been established.

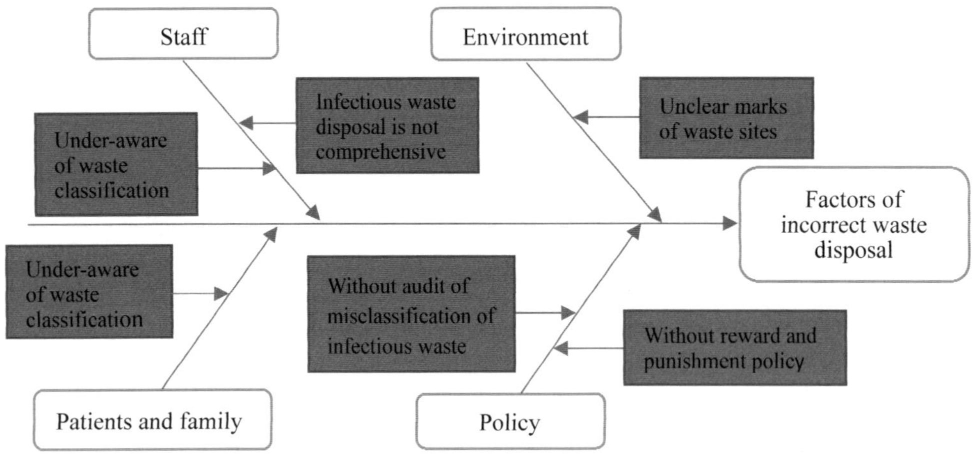

Figure 1. Cause & Effect diagram of the incorrect disposal of medical waste

2.5. Improvement strategies

2.5.1. Environment aspect. Previous infectious waste collection spots in hospital were not clearly marked for infectious waste, resulting in classification errors in the handling of medical waste by staff or the family of the patients. The proposed solution is to place physical photographs next to trash buckets where infectious waste is placed to assist staff and patients' families in discarding infectious waste.

2.5.2. Policy aspect. The analysis of the medical waste treatment issues (in Figure 1) shows that the hospital has not yet formulated a reward and punishment policy for the treatment of medical waste. The corresponding proposal is to hold a competition "medical waste reduction and classification" and also offer rewards. In addition, a "medical waste classification audit team" was set up by 5S (Seiri, Seiton, Seiso, Seiketsu, Siesuke) audit members to conduct occasional spot checks on the classification of medical waste in each area of responsibility and to continuously assess performance [9].

2.5.3. Employees aspect. The staff of this branch are under-aware of the classification of medical waste, and the corresponding proposal is to carry out "medical waste classification and collection" education and training. In the case of the inaccuracy of the disposal of infectious waste, the proposed solution is to add "waste liquid collection tanks" to the care vehicles.

2.5.4. Patients and family members aspect. The patients and family members are also under-aware of the medical waste classification. The proposed solution is to strengthen the environmental introduction of admission, point out the location of the medical waste disposal rooms, and set up "medical waste disposal locations and direction maps" at the nursing station to assist the patients and their families to find the correct waste placement site. It also strengthens the importance of publicizing medical waste classification for patients and their families.

3. Results and analysis

3.1. Questionnaire

In order to examine the proposed alternatives and the effect of the implementation, the Quality Assurance Team of administration office designed a questionnaire for patients and family members and nursing staff respectively.

3.1.1. Patient and family questionnaire

(1) Were contaminants collected in the nursing station?
(2) Were the diapers thrown into the red trash buckets?
(3) Were cotton balls and cotton sticks thrown into the blue trash buckets?
(4) Were wipe paper and toilet paper thrown into the blue trash buckets?
(5) Were Styrofoam boxes thrown into the red trash buckets?
(6) Were flowers thrown into the blue trash buckets?
(7) Were masks thrown into the blue trash buckets?
(8) Were newspapers thrown into the recycling bins?
(9) Were polyethylene bottles thrown into the recycling bins?
(10) Was the kitchen waste thrown into the red trash buckets?

In response to the above-mentioned questionnaire, a survey was given before and after the medical waste publicity and education activities were conducted, and 18 patients and family members were interviewed before the event, including 8 men and 10 women, with a correct response rate of 74.4%. After the event, a total of 21 patients and family members were interviewed, including 6 men and 16 women, and the accuracy rate of answering the questionnaire was 93.3%, an increase of 25.4%.

3.1.2. Nursing staff questionnaire

(1) Were cotton sticks that are not contaminated with the blood and body fluids of infected people thrown into the red trash buckets?
(2) Were pots, urinals, and diapers used by non-infected patients thrown into the red trash buckets?

(3) Was medical waste after chemotherapy thrown into the red trash buckets?
(4) Were drain fluids in drainage bottles (bags) thrown into the red trash buckets?
(5) Were general Amp and Vial within 50ml thrown into the blue trash buckets?
(6) Were IV sets thrown into the blue recycling bins?
(7) Were general IV infusion bags after removing moisture and stickers thrown into the blue recycling bins?
(8) Were PVC empty bottles of preparations treated as general waste?
(9) Were polyethylene bottles and aluminum cans thrown into the yellow trash buckets?
(10) Was medical waste divided into general and infectious waste?

The accuracy rate of responses to the above-mentioned questionnaire by nursing staff before education and training was 84.3%, and the accuracy rate of answers after education and training was 95.8%, an increase of 11.5%.

3.2. Results of implementation

3.2.1. The statistics of medical waste amount. Ward waste measured 0.54 kg per bed per day before education and training activities, and 0.33 kg per bed per day after education and training activities. As a result, the daily amount of waste per bed decreased from 0.54 kg to 0.33 kg, a decrease rate of 38.9% (see Table 3).

Table 3. Statistics on infectious waste in wards

	Before improvement actions (year B/11/01~year B/11/07)	After improvement actions (year C[a]/01/18~year C/01/23)
Total weight(kg)	150	**76**
Total bed-day amount	280	**231**
Average weight per bed (kg/bed-day)	0.54	**0.33**

[a]Note: C=B+1

3.2.2. Waste disposal cost

Monthly saving: \$26.5 /kg * 1,271bed-day * (0.54-0.33) kg/bed-day = \$7,073/month
Yearly saving: \$84,876/year

3.2.3. Differences between before and after improvement actions

(1) Increased awareness of medical waste classification among patients and family members, with a 25.4% increase in test scores.
(2) Increased awareness of medical waste classification among nursing staff, with test scores increasing by 11.5%.
(3) Reduced the amount of infectious waste in the wards from 0.54kg/bed-day to 0.33kg/bed-day, with a reduction rate of 38.9%.
(4) Increased the number of illustrated guidelines in wards and medical waste disposal rooms from 0 to 20.
(5) Increased the number of posters for the classification of waste in wards from 1 to 21.

4.Conclusion

After the education and publicizing of medical waste identification to medical staff, patients and family members in the Jiayi Branch, the awareness of nursing staff, patients and their families regarding medical waste was greatly improved, directly reducing the amount of infectious waste and reducing the cost of its treatment. As Gama Hospital has several branches in Taiwan, the hospital can promote a more comprehensive approach to reduce secondary pollution caused by improper handling of medical waste, so as to fulfill the social responsibility of environmental sustainability.

IOP Conf. Series: Earth and Environmental Science **728** (2021) 012004 doi:10.1088/1755-1315/728/1/012004

5.References

[1] World Health Organization 2018 Medical waste http://who.int/zh/news_room/fact_sheets/detail/health-care-waste
[2] Pe´pin J, Abou Chakra CN, Pe´pin E, Nault V, Valiquette L. 9 June 2014 Evolution of the global burden of viral infections from unsafe medical injections 2000-2010 *PLOSOne* **9(6)**: e99677
[3] Arterial Network 28 August 2018 How to dispose of 2 million tons of medical waste a year? http://m.huxiu.com/article/259830.html
[4] Environmental Protection Bureau of the Executive Council 2006 Medical waste disposal methods http://oaout.epa.gov.tw/law/LawContent.aspx?id=GL003544
[5] Chen W 2016 Biomedical waste management and disposal http://gao.sinica.edu.tw/ehsmd/ch/docu/education/20160824_sinica_Biowaste_print_pdf
[6] Lan Z 2000 *Infectious medical waste treatment strategy and technology review* Policy Research Foundation
[7] Cai Q 2001 *The final report of the study of medical waste identification, classification and treatment guidelines* Published by Environmental Protection Bureau of the Executive Council
[8] *Review and regulations of the determination of the characteristics of infectious industrial waste* 1998 Published by Environmental Protection Bureau of the Executive Council
[9] 5S-Seiri, Seiton, Seiso, Seiketsu, Shitsuke 17 November 2016 http://ncl.com.tw/world/250

Influence of Temperature and Reaction Time on the Performance of Thermal Pyrolysis of Compact Polystyrene Waste.

***P Palmay[1], R Nuñez[2], C Donoso[3] and J Bruno[4]**

1, 2 Facultad de Ciencias, Escuela Superior Politécnica de Chimborazo ESPOCH, Panamericana Sur Km 1 1/2, Riobamba (Ecuador)
3 Departamento de Ciencias de la Energía y Mecánica, Universidad de las Fuerzas Armadas ESPE, Av. General Rumiñahui s/n Sangolqui, P.O. Box171-5.231, Ecuador
4 Universitat Rovira i Virgili, Dept. of Mechanical Engineering, Avda. Paisos Catalans, 26, 43007-Tarragona (Spain)
*E - mail: paul.palmay@espoch.edu.ec

Abstract. The purpose of this study is to determine the influence of the temperature and reaction time on the thermal pyrolysis process of compact polystyrene residues based on the percentage of liquid obtained at a heating cup of 15 C min^{-1} and cooling system for the condensation of pyrolytic gases at 10 ° C. It worked with temperatures between 375 and 425 ºC and times of 5 and 10 minutes. The results showed that the reaction time has no representative influence on the liquid fraction, while the temperature has a great importance in the liquid yield reaching a maximum contribution at 400 ° C, after this the percentage of liquid oils tends to decrease by guiding the formation of the gas fraction. The maximum production was 81.4% with a kinematic viscosity of 1,026 $mm^2\ s^{-1}$, relative density 0.9352, flash point 24 ºC and a caloric power of 48.5 MJ kg^{-1}. The mixture of aromatic hydrocarbons obtained, with carbon chains between C6 and C20, are present in 75% by mass.

1. Introduction

The production of plastics is increasing rapidly in recent years due to its high demand, versatility, low cost and its use in a number of industrial and commercial applications giving counterproductive environmental result by the generation of waste, which is one of the fractions of greater attention in municipal solid waste (MSW) for its long time of degradation and high volume used for final disposal [1]. In the MSW exist a mixture of plastics that can be considered as recyclable that constitute several thermoplastics such as PET polyethylene terephthalate, PP polypropylene, PE polyethylene, PS polystyrene, among others. Plastic waste is managed through various techniques including recycling, reuse, reduction of use and handling - disposal in landfills as conventional techniques or non-conventional techniques such as chemical recycling [2]. In recent years chemical recycling has become one of the most important methods in several countries with techniques such as glycolysis, gasification, hydrolysis, among others to convert waste into fuels [3], technologies such as gasification and pyrolysis have been the subject of researches in recent years. [4].
Pyrolysis is a process of converting organic material including plastic waste into energy (liquid oils) and value-added products (coal) by the controlled addition of heat with or without catalytic agents that guide the formation of compounds similar to conventional fuels [5]. Polystyrene pyrolysis is mainly performed as a tool for depolymerization of macromolecule strictly controlled conditions and

subsequent recovery of styrene monomer of great commercial interest. Furthermore, if a complete pyrolysis of the polystyrene plastic molecule is carried out, the breaking of the chains into lighter fractions [6], which follows a reaction mechanism in a single step for most plastics except PVC, data verified by means of the DTG curve in the thermogravimetric analysis where there is only one peak in the derivative of the mass versus the temperature with respect to time [7]. The objective of this article is to identify the temperature and time of greater production of liquids in the thermal pyrolysis of compact polystyrene, a plastic used mainly in the food industry as a container; and the characterization of said product according to its chemical composition and physical-chemical properties.

2. Materials and methods

2.1. Sample preparation and reactor conditions

A thermal pyrolysis reactor designed and built for the effect was used, of batch type with heating by electrical resistances and a system of valves to purge with a continuous flow of nitrogen the present. The heating cup is controlled through the temperature control system. The gases coming from the reaction are cooled in a system with cooling water at 10 °C. The reactor pressure at the time the plastic degradation begins is maintained at pressures below 550 kPa for safety and control of the equipment, which is heated to a fixed rate of 15 °C min^{-1}, a common speed in a slow pyrolysis where the effect of other variables can be evaluated. For each test, 100 grams of sample were weighed and 4 temperature levels (375, 400, 425 and 450 °C) were carried out to determine the effect of this variable on the production of the liquid fraction, at two different residence times that were 5 and 10 minutes once the working temperature was reached, with three repetitions by experiments to ensure its repetitiveness in the different levels and treatment, having a set of 24 data used to study the influence of the two input variables on the amount of liquid generated, which will determine the percentage of the fraction at the end of each of the repetitions by weighing it and then by means of a mass balance according to the plastic fed.

2.2. Analytical methods

The polystyrene residues were analyzed by thermogravimetry (TGA 1 STAR System METTLER TOLEDO) establishing the limits of temperature of the experimentation that served in the experimental design established in a range from 25 to 900 °C with neutral atmosphere of nitrogen and a heating rate of 15°C/min. The pyrolytic products are condensed at a temperature of 10°C; where the non-condensable gas fraction is collected in Tedslar casings through a gas trap and the liquid fraction sampled in amber containers and maintained at temperatures of 15°C. At least three repetitions of the experiment were performed to obtain a significant sample size and to analyze the effect of temperature and time variables. Temperature and pressure control is performed by taking data every thirty seconds to determine the tendency of gas production at the time of increasing the temperature Figure 1.

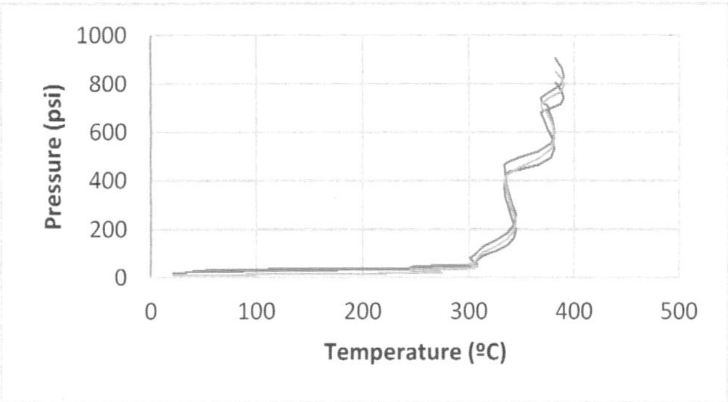

Figure 1. Reactor temperature versus pressure produced by the gases generated.

3. Results and Discussion

3.1. Raw material analysis

The recycling of polystyrene is rare in recycling campaigns due to its difficult identification concerning waste from other types of plastics, which is why it is necessary to characterize the material sampled. In addition to the physical tests of solubility and flame color, the FTIR of the compact PS was carried out, where four groups of pronounced bands belonging to the multiple tension movements are presented: a) C-H bonds to 2919,7cm^{-1}, b) C-C bond between 1400, 1600cm^{-1} aromatic ring with -CH2 bending movement, c) aromatic ring tensions with values of 748,245cm^{-1} and d) ring deformation vibrations outside the plane at a length of 1027 cm-1, as shown in Figure 2; confirming the majority presence of styrene in the structure of the polymer [8], [9]. The band present in 1341 cm^{-1} shows the presence of N-C and N-H bonds corresponding to amines and amides that are UV protective additives or antistatic additives, very common in the manufacture of plastics to give better characteristics. The band in 1027 cm^{-1} corresponding to the vibration of methylbenzene bonds that are characteristic of plasticizing agents.

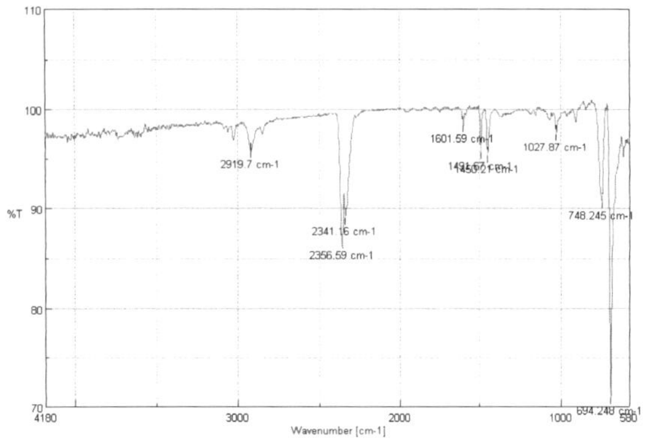

Figure 2. Infrared Spectroscopy Results.

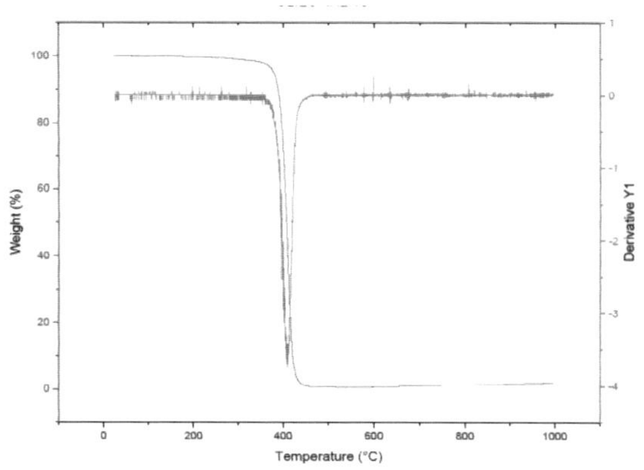

Figure 3. TGA Y DTG of the Polystyrene

In order to determine the thermal degradation temperature that provides the best performance of the liquid fraction, the TGA and DTG curve in Figure 3 establishes the initial degradation temperature for PS at 370ºC and a maximum temperature of 425ºC, worked under dynamic conditions with a heating rate of 15ºC, which compared with data shown by [10] verifies that the displacement of the TGA curves to the high temperature region when the heating speed increases. That is, at a faster heating rate, the equilibrium is reached slowly and, as such, the curve moves towards the high temperature region.

3.2. Pyrolysis performance

During the first stage of thermal degradation of PS a cracking of volatile substances occurs at relatively low temperatures (<150ºC) due to the rupture of bonds in the structures as additives or contaminating agents of the material, followed by a series of parallel decomposition reactions such as reforming, dehydration, condensation, polymerization, oxidation and gasification [11], The performance averages of pyrolytic products with a constant heating rate of 15 ºC min^{-1} are shown in the table 1.

The elemental analysis of the PS is characterized by the low ash content producing a low percentage of solid product after pyrolysis at different temperatures [12], a reference that explains the values lower than 3% in the experiments of the solid fraction. At temperatures close to 375°C there is a favourable effect of this parameter on the performance of the liquid fraction when the residence time increases from 5 to 10 minutes, however as the temperature increases to 450°C the residence time has no significant effect on any of the product fractions. In similar studies carried out at 450°C and heating rates lower than 10°C per minute, a fraction similar to that of this research was obtained in the liquid product, i.e. 80%, however the gas fraction decreases to 13% and the percentage of solids increases to 6.2% [4]; this is possibly attributed to the regrouping of compounds to form long chains of higher molecular weight, when the residence time increases.

Table 1. Pyrolysis results of Polystyrene at different time and temperature treatments.

T (°C)	t(min)	% LIQ	% SOL	%GAS	t(min)	% LIQ	% SOL	%GAS
375	5	73,79	1,84	24,37	10	76,56	0,90	22,54
400		80,60	0,56	18,84		81,41	1,16	17,43
425		78,35	0,64	21,01		79,08	2,28	18,64
450		80,63	2,16	17,21		79,28	2,30	18,42

3.3. Effect of reaction time on percentage of liquid.
The effect of time at four temperature levels was studied. At a time of 5 minutes at 400°C it gives a yield of 80.6% and at 10 minutes a yield of 81.41% showing that there is no significant difference in the production of liquids at the reaction times of 5 minutes and 10 minutes in the reactor conditions. The amount of total hydrocarbons in the 5-minute sample presented a greater dispersion of data due to the fact that the effect of time is a function of the dimensions of the reactor and the transfer of heat through the elements of the equipment. Therefore a time of 10 minutes is suggested as an optimal time in which the distribution of hydrocarbons in the sample is stabilized, this being the time necessary for the total degradation of the PS. Similar analysis was observed and reported by [4] and [13].

3.4. Effect of temperature on percentage of liquid.
The quantity of gas is closely related to the quantity of liquid pyrolysis products, from 425 °C the liquid fraction begins to decrease to favour the gaseous fraction [14]. A high temperature causes the C-C bonds to have a greater rupture generating lighter hydrocarbons with a short chain, behaviour explained by [15]. Thus, for every 1% that the liquid fraction decreases, the fraction of gas generated increases by 0.97%.
The different temperature levels at a time of 10 minutes in the batch reactor gave the yields presented in Figure 4. At that time at a temperature of 375°C gave a liquid fraction of 76.6% with a gas production of 22.5% and 0.9% solids which was the lowest yield obtained. The highest production value was at 400 °C with 81.4%, 17.4% and 1.2% of liquid, gaseous and solid fraction respectively.
It is determined as the optimum temperature of greater production of liquid percentage at the operating conditions defined above, 400 °C; different to that predicted by the thermogravimetric analysis (450°C) due to the fact that in the TGA there are controlled conditions with small amounts of sample, which in comparison with the conditions of the pyrolysis reactors can be, even in occasions, particular for each case [4].
The yield of the produced liquid oil rose at the beginning of the process, after a certain temperature and time, the yield of the oil began to decrease (Figure 4). This is due to the fact that after reaching the temperature of maximum generation of liquid compounds certain secondary reactions such as poly aromatic formation reactions are initiated during the pyrolytic process which decreases the performance of oils [16] and [17].

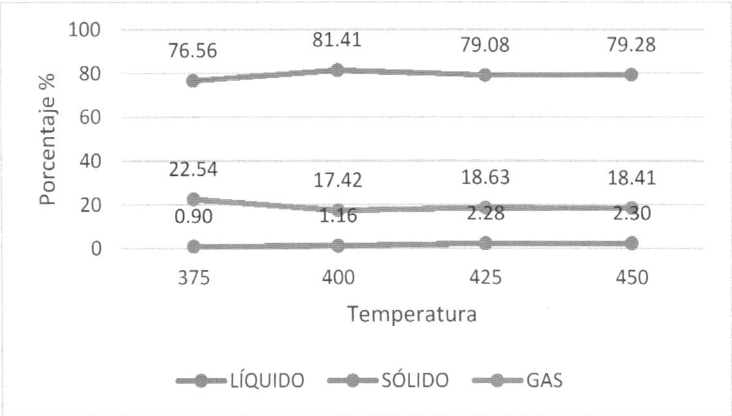

Figure 4. Distribution of the liquid, solid and gaseous pyrolysis fractions of PS

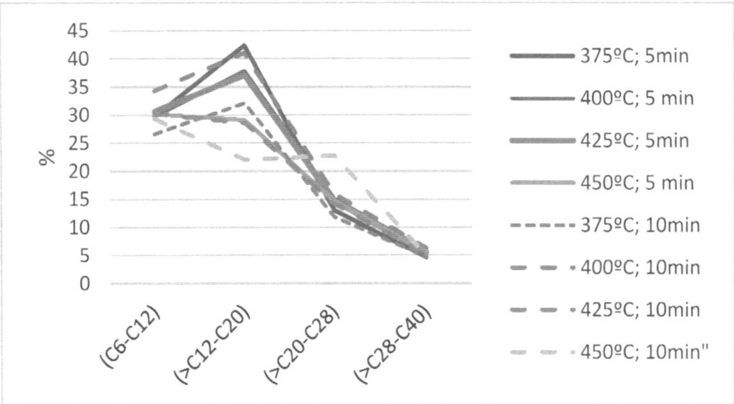

Figure 5. Distribution of the composition of the liquid fraction at 5 and 10 minutes of reaction.

The pyrolysis of PS at a temperature of 400°C and residence time of 10 min, produced 41% of carbon chain components between C12 and C 20. Compounds that are formed possibly due to the reactions that occur at the time of production of styrene which is the hydrocarbon that is first generated in the decomposition of PS [18]. As can be seen in Figure 5 up to 425°C a similar behavior is seen in the production of light compounds, however at 450°C a greater production of compounds between C20 and C28 is observed as opposed to the other temperatures. Particularly evident at both 5 and 10 minutes of reaction but much more pronounced at 10 minutes congruent with [4] where it mentions that the longer the residence time facilitates that the chains again interlace and form linear compounds of higher molecular weight. Previous studies have reported an increase in the generation of styrene at higher temperatures up to a maximum point, with a subsequent decrease due to the formation of compounds originating between several aromatic radicals [19].

3.5. Characteristics of the liquid obtained

Different parameters of the liquid fraction obtained for the best pyrolysis conditions were analyzed, as shown in table 2. The kinematic viscosity presents a value of 1,026 $mm^2\ s^{-1}$ at 40°C, which is lower than that specified in the ASTM D975 standard for conventional diesel, however it is consistent with the value reported by [4]. The low viscosity of the liquid fraction is attributed to the variation in the structure and composition of PS that differentiates it from other monomers [20]. A low viscosity favors the pulverization and atomization of the fuel, contributing to engine performance, even at low temperatures [21]. The relative density at 15.6°C of the liquid fraction obtained is 0.9352 $g\ cm^{-3}$, which resembles the values reported by [4]; this density is slightly higher than that reported for conventional diesel (0.81-0.87 g cm^{-3}), as reported by [22]. The sulfur content of the liquid fraction obtained is extremely high, 1120 ppm, which means that for its use as fuel must first undergo a previous desulfurization process, however, when applied as an additive to reduce viscosity, for example the cutter stock, the sulphur content is within permissible limits. The HHV of the liquid fraction obtained at the best pyrolysis conditions is one of the most important parameters for the identification of a fuel, the value obtained was 48.5568 MJ kg^{-1}, which is higher than the reference value for conventional diesel (43.06 MJ kg^{-1}) [23]. Several researchers have studied the liquid fraction of PS pyrolysis as an individual energy source or as a mixture with conventional diesel [24], due to the high content of aromatic compounds that raise the flash point. However, there are other applications in which the main objective is the recovery of monomers mainly styrene, and other compounds such as toluene and ethylene benzene that can be used as a chemical source in industries for the polymerization of PS polymer [15].

Table 2. Characteristics of the liquid oils obtained.

TEST	UNITS	RESULTS		CUTTER STOCK PARAMETERS
		Authors	Miandad (2016)	
API Gravity at 60 °F	°API	19,8	NR	NR
Relative Density 15,6°C.	-	0,9352	0,91 - 0,92	NR
Sulfur content	% W	0,112	NR	Max 2,5
Kinematic viscosity at 40°C	$mm^2\ s^{-1}$	1,026	1,92 – 3,24	Max 7
Water and sediment content	%V	0,5	NR	Max 0,2
Distillation of 90%	°C	325	NR	NR
Flash point	°C	24,1	28,1 – 30,2	Min 60
Calculated cetane number	-	20,06	NR	NR
***Higher Heating Value* (HHV)**	MJ/kg	48,5568	41,4 – 41,8	NR

*NR: Not Reported

4. Conclusions

The results of obtaining liquid hydrocarbons from PS in a 3 L capacity batch reactor with a coupled cooling system for the condensation of pyrolytic gases at 10ºC were studied. In which the effect of time and temperature variables on the performance of liquid products was analysed. The reaction time taken since it reaches the working temperature has no influence and is worked at 10 minutes. The temperature has high significance in the liquid yield until reaching a maximum contribution of 400 ºC after this the percentage of liquid oils decreases as the temperature increases generating an increase in the formation of the gas fraction. 81.4% of liquid products are obtained with a kinematic viscosity of 1,026 mm2 s^{-1}, relative density 0.9352, flash point 24 ºC and a caloric power 48.5 MJ kg^{-1}.

Due to the high content of styrene, toluene and ethylene benzene in pyrolytic oils according to [16] and [15]. The monomer generated can be used in the chemical industry [25]; [4] or for its high content of aromatic hydrocarbons and mixture of carbon atoms between C9 and C25 and flash point lower than 100ºC [26]; as well as for the chromatographic analysis that shows a composition of C6 to C20 of 75%, the fuel obtained from the liquid fraction has composition and properties similar to those of the cutter stock and can be used as diluent in refineries to adjust the viscosity of fuel oil.

5. References

[1] Gourmelon, G. (2015). Global plastic production rises, recycling lags. New Worldwatch Institute analysis explores trends in global plastic consumption and recycling. Recuperado de http://www. worldwatch. org, 208.

[2] Ouda, O. K., Raza, S. A., Nizami, A. S., Rehan, M., Al-Waked, R., & Korres, N. E. (2016). Waste to energy potential: a case study of Saudi Arabia. *Renewable and Sustainable Energy Reviews*, *61*, 328-340.

[3] Nizami, A., Rehan, M., Ouda, O. K., Shahzad, K., Sadef, Y., Iqbal, T., & Ismail, I. M. (2015). An argument for developing waste-to-energy technologies in Saudi Arabia. *Chemical Engineering Transactions*, *45*, 337-342.

[4] Miandad, R., Rehan, M., Nizami, A. S., Barakat, M. A. E. F., & Ismail, I. M. (2016). The energy and value-added products from pyrolysis of waste plastics. In *Recycling of solid waste for biofuels and bio-chemicals* (pp. 333-355). Springer, Singapore.

[5] Rathore, D., Nizami, A. S., Singh, A., & Pant, D. (2016). Key issues in estimating energy and greenhouse gas savings of biofuels: challenges and perspectives. *Biofuel Research Journal*, *3*(2), 380-393.

[6] Waggoner, D. C., & Hatcher, P. G. (2016). Thermally assisted methylation of glycerol by tetramethylammonium hydroxide thermochemolysis. *Journal of analytical and applied pyrolysis*, *122*, 289-293.

[7] Grammelis, P., Basinas, P., Malliopoulou, A., & Sakellaropoulos, G. (2009). Pyrolysis kinetics and combustion characteristics of waste recovered fuels. *Fuel*, *88*(1), 195-205.

[8] Asensio, R. C., Moya, M. S. A., de la Roja, J. M., & Gómez, M. (2009). Analytical characterization of polymers used in conservation and restoration by ATR-FTIR spectroscopy. *Analytical and bioanalytical chemistry*, *395*(7), 2081-2096.

[9] Miandad, R., Barakat, M. A., Aburiazaiza, A. S., Rehan, M., Ismail, I. M. I., & Nizami, A. S. (2017). Effect of plastic waste types on pyrolysis liquid oil. *International biodeterioration & biodegradation*, *119*, 239-252.

[10] Nisar, J., Ali, G., Shah, A., Iqbal, M., Khan, R. A., Anwar, F., ... & Akhter, M. S. (2019). Fuel production from waste polystyrene via pyrolysis: Kinetics and products distribution. *Waste management*, *88*, 236-247.

[11] Sogancioglu, M., Yel, E., & Ahmetli, G. (2017). Investigation of the Effect of Polystyrene (PS) Waste Washing Process and Pyrolysis Temperature on (PS) Pyrolysis Product Quality. *Energy Procedia*, *118*, 189-194.

[12] Wu, J., Chen, T., Luo, X., Han, D., Wang, Z., & Wu, J. (2014). TG/FTIR analysis on co-pyrolysis behavior of PE, PVC and PS. *Waste management*, *34*(3), 676-682.

[13] Lopez, A., De Marco, I., Caballero, B. M., Laresgoiti, M. F., & Adrados, A. (2011). Influence of time and temperature on pyrolysis of plastic wastes in a semi-batch reactor. *Chemical Engineering Journal*, *173*(1), 62-71.

[14] Zhou, H., Wu, C., Onwudili, J. A., Meng, A., Zhang, Y., & Williams, P. T. (2016). Influence of process conditions on the formation of 2–4 ring polycyclic aromatic hydrocarbons from the pyrolysis of polyvinyl chloride. *Fuel Processing Technology*, *144*, 299-304.

[15] Artetxe, M., Lopez, G., Amutio, M., Barbarias, I., Arregi, A., Aguado, R., ... & Olazar, M. (2015). Styrene recovery from polystyrene by flash pyrolysis in a conical spouted bed reactor. *Waste management*, *45*, 126-133.

[16] Jung, S. H., Kim, S. J., & Kim, J. S. (2013). The influence of reaction parameters on characteristics of pyrolysis oils from waste high impact polystyrene and acrylonitrile–butadiene–styrene using a fluidized bed reactor. *Fuel processing technology*, *116*, 123-129.

[17] Mo, Y., Zhao, L., Wang, Z., Chen, C. L., Tan, G. Y. A., & Wang, J. Y. (2014). Enhanced styrene recovery from waste polystyrene pyrolysis using response surface methodology coupled with Box–Behnken design. *Waste management*, *34*(4), 763-769.

[18] Onwudili, J. A., Insura, N., & Williams, P. T. (2009). Composition of products from the pyrolysis of polyethylene and polystyrene in a closed batch reactor: Effects of temperature and residence time. *Journal of Analytical and Applied Pyrolysis*, *86*(2), 293-303.

[19] Demirbas, A. (2004). Pyrolysis of municipal plastic wastes for recovery of gasoline-range hydrocarbons. *Journal of Analytical and Applied Pyrolysis*, *72*(1), 97-102.

[20] Siddiqui, M. N., & Redhwi, H. H. (2009). Pyrolysis of mixed plastics for the recovery of useful products. *Fuel Processing Technology*, *90*(4), 545-552.

[21] Mohan, D., Pittman Jr, C. U., & Steele, P. H. (2006). Pyrolysis of wood/biomass for bio-oil: a critical review. *Energy & fuels*, *20*(3), 848-889.

[22] Syamsiro, M., Saptoadi, H., Norsujianto, T., Noviasri, P., Cheng, S., Alimuddin, Z., & Yoshikawa, K. (2014). Fuel oil production from municipal plastic wastes in sequential pyrolysis and catalytic reforming reactors. *Energy Procedia*, *47*, 180-188.

[23] Kamal, D. M., & Zainuri, F. (2015). Green product of liquid fuel from plastic waste by pyrolysis at 900 C. *Journal of Energy and Power Engineering*, *9*, 40-44.

[24] Lee, S., Yoshida, K., & Yoshikawa, K. (2015). Application of waste plastic pyrolysis oil in a direct injection diesel engine: for a small scale non-grid electrification. *Energy and Environment Research*, *5*(1), 18.

[25] Achilias, D. S. (2007). A review of modeling of diffusion controlled polymerization reactions. *Macromolecular theory and simulations*, *16*(4), 319-347.

[26] Méndez, Y., & Tancredo, E. (2017). *Estudio de la estabilidad intrínseca de mezclas de fuel oil base y cutter stock para la obtención del fuel oil comercial* (Bachelor's thesis, Quito: UCE).

IOP Conf. Series: Earth and Environmental Science **728** (2021) 012006 doi:10.1088/1755-1315/728/1/012006

Copper, Nickel, Lead, and Cadmium Concentration in Soils around Industrial Areas from Latacunga, Ecuador

Jessenia Cela, William Bonilla and Ricardo Urrutia-Goyes

Departamento de Ciencias de la Energía y Mecánica, Universidad de las Fuerzas Armadas ESPE, Av. Gral. Rumiñahui s/n, P.O. Box 171-5-231B, Sangolqui 171103, Ecuador
E-mail: erurrutia@espe.edu.ec

Abstract. Environmental pollution has increased in recent decades as a result of various human activities and the presence of various chemical compounds, including heavy metals. The goal of this paper is to determine the concentration and distribution of copper, nickel, lead, and cadmium in soils around industrial areas from Latacunga, Ecuador. Soil samples were obtained from 19 localities: two zones with influence from the industrial sector and two recreation areas and public access. Samples obtained were pretreated and then analyzed by flame atomic absorption spectrophotometry following USEPA 3050B and USEPA 3052 methodology. Finally, a statistical analysis belonging to a randomized complete block design, an analysis of variance (ANOVA) was applied (confidence level of 95%). The results show that the concentration of the metals analyzed decreased in the order of nickel> copper> lead. Additionally, is it shown that the concentration of nickel exceeds the limits suggested by local regulators in all study zones. On the other hand, Zone 1 and Zone 2 show copper and nickel concentrations that exceed the referenced limits.

1. Introduction

The contamination of water, air, and soil has been among the main environmental problems worldwide for decades. Some instances of said contamination come from human activities and bad practices related to the improper treatment of wastes. Pollution can be happen then, due to the presence of various waste chemical compounds in environmental media, including heavy metals, causing a threat to public health and environmental safety. Heavy metals, when found in soil, can be transferred to living organisms through the food chain [1]. Heavy metals, however, are present in the earth's crust in elemental form or as part of other chemical compounds. Their concentration vary depending on the type of soil and the activity that is carried out [2] since the natural concentration of a metal in soil derives from the bedrock that originated the soil, while the amount of metal that is incorporated into the soil will depend on the degree of resistance of the mineral that constitutes it and anthropogenic actions [2].

The presence of heavy metals in soil can become a contamination problem when concentrations become representative and even excessive. Several studies regarding the presence of heavy metals in air, water, and sediments have been proposed around the globe, however, there are few studies in Ecuador regarding this matter [3][4]. The city of Latacunga, in the Andean region of Ecuador, is the capital of the province of Cotopaxi at an altitude of approx. 2,850 meters above sea level where cold weather predominates. The area is purely volcanic, and five types of soils can be distinguished i.e. entisol, histosol, inceptisol, mollisol and untransformed volcanic remains [5].

Within the city of Latacunga, there are green areas with public access and green areas in contact with wastes generated by different kind of industries. The main objective of this paper is to analyze the

IOP Conf. Series: Earth and Environmental Science **728** (2021) 012006 doi:10.1088/1755-1315/728/1/012006

concentration of copper, nickel, lead and cadmium in soil samples from green areas with public access and green areas with influence from effluents generated by a foundry, a food (flour) industry, and a textile industry in the city of Latacunga. Finally, areas with concentrations that exceed the maximum limits established in the country are determined.

2. Methodology

The city of Latacunga is located at an altitude of approx. 2,850 meters above sea level. Weather is predominately cold and low in humidity. The geographical extension covers around 138,630 hectares with diverse soil formations throughout its territory [5]. For the development of this study, three green areas with public access and two areas with direct influence of two types of industries belonging to the city center were taken into account, as well as one area with direct influence by an industry in the rural area of the city as can be seen in Figure 1.

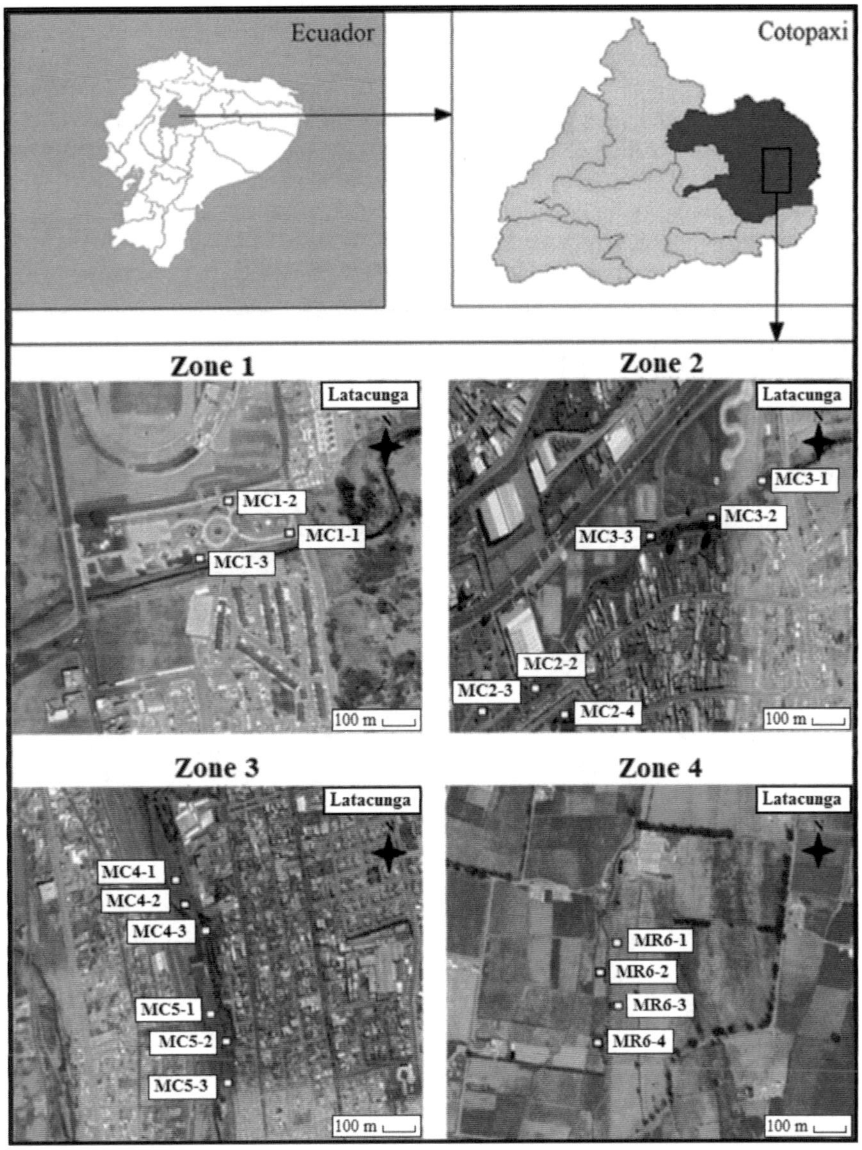

Figure 1. Study area in the city of Latacunga. 19 sampling points located in the city center (Zone 1, Zone 2 and Zone 3) and a rural sector (Zone 4). Basemap data © Google 2019.

2.1. Sampling
Soil samples collected for this research were obtained during February 2019. Samples were obtained from a total of 19 points, establishing 6 collection areas: three of them have direct influence from the industrial sector and three are recreational areas and are of public access. At each sampling point, the approx. amount of 200 g of surface soil was collected at a depth between 10 and 20 cm with a plastic spatula after vegetable remains were removed. Each soil sample was stored in plastic containers and covered with a cloth to prevent conductivity or oxidation of the metals present in the soil. Once the samples were collected and stored with their respective labeling, they were dried in the laboratory at room temperature for a period of 96 hours and then dried in an oven at 60°C for 72 hours [6].

2.2. Analytical procedures
The nineteen soil samples were subjected to a 38 μm plastic sieve and 6 random samples (one sample for each collection area) were subjected to a 1,000 μm sieve, totaling 25 samples for chemistry and composition determination, in order to analyze the variability related to metal concentration.
The chemical treatment for the digestion process of the samples was based on USEPA 3050B [7] and USEPA 3052 standards [8]. Briefly, each sample had to be homogenized and sieved. An amount of 0.1 g of the sample was then added to 8 ml of HNO_3, 5 ml of HCl, 1 ml of HF and 5 ml of concentrated H_3BO_3. Under operating conditions in a microwave digestion equipment (ETHOS ONE), the added acids were allowed to act to subsequently obtain a solution that was diluted with distilled water to 100 ml. Finally, the resulting solution was used for the analysis by using atomic absorption spectroscopy.
The soil samples obtained from the digestion process were analyzed by using a Perkin Elmer flame atomic absorption spectrophotometer (model Aanalyst 300). Copper, nickel, lead and cadmium were determined. The equipment was calibrated with deionized water and standard solutions of 5, 15, and 30 ppm for each element analyzed in order to prepare calibration curves.
Chemical analysis was performed using deionized water, distilled water, 65% HNO_3, 37% HCl, 40% HF, and 5% H_3BO_3. The standard solutions used were typical of the atomic absorption equipment. Instrumental analysis was carried out using a Digestion Microwave (ETHOS ONE) and a Perkin Elmer (model Aanalyst 300) flame atomic absorption spectroscopy kit.

2.3. Statistical analysis
The concentration data were processed with statistical software package Minitab. Descriptive statistics were obtained and the analysis was performed using a randomized complete block design, with a confidence level of 95% ($\alpha = 0.05$).

3. Results and discussion
Soil samples information was contrasted with the maximum permissible limits established in Ecuadorian standards, where values of 0.5 mg/Kg for cadmium, 30 mg/Kg for copper, 20 mg/Kg for nickel and 25 mg/Kg for lead are mentioned [9].

3.1. Heavy metal concentration in soil samples by zone
A comparative analysis between zones was carried out with respect to the concentration of metals, where sampling points were grouped. Zone 1 refers to samples obtained at La Cocha sector (MC1, see Fig. 1), Zone 2 refers to samples collected at Ignacio Flores Nautical Park and Illuchi River banks (MC2 and MC3), Zone 3 refers to the banks of Cutuchi River around a food industry and an aluminum foundry (MC4 and MC5), and Zone 4 refers to Saquimala River banks close to a textile industry in a rural area (MR6).
Regarding the concentration of copper, nickel, lead and cadmium in the study areas, the highest copper concentrations were obtained in Zone 1. Nickel concentrations were higher in Zone 1 and Zone 3. Lead concentrations were the highest in Zone 2. Cadmium concentrations in all zones were below the minimum detection value of 0.1 mg/Kg. Metal concentrations analyzed in Zone 1 decreased in the order copper > nickel > lead. In Zone 2, they decreased in the order copper > lead > nickel. In Zone 3, they decreased in the order nickel > copper > lead. In Zone 4, it followed the order nickel > copper > lead. Concentrations of heavy metals in the study areas can be seen in Figure 2.

As shown in Table 1, concentration values for copper in Zone 1 exceed the permissible reference values, [9] but not in the rest of the zones. With respect to nickel, it was found that all the studied areas exceed the concentration values compared to the permissible reference values, while lead concentration values exceed the permissible reference values only in Zone 2. Finally, cadmium concentration values were always below the minimum detection value of 0.1 mg/Kg.

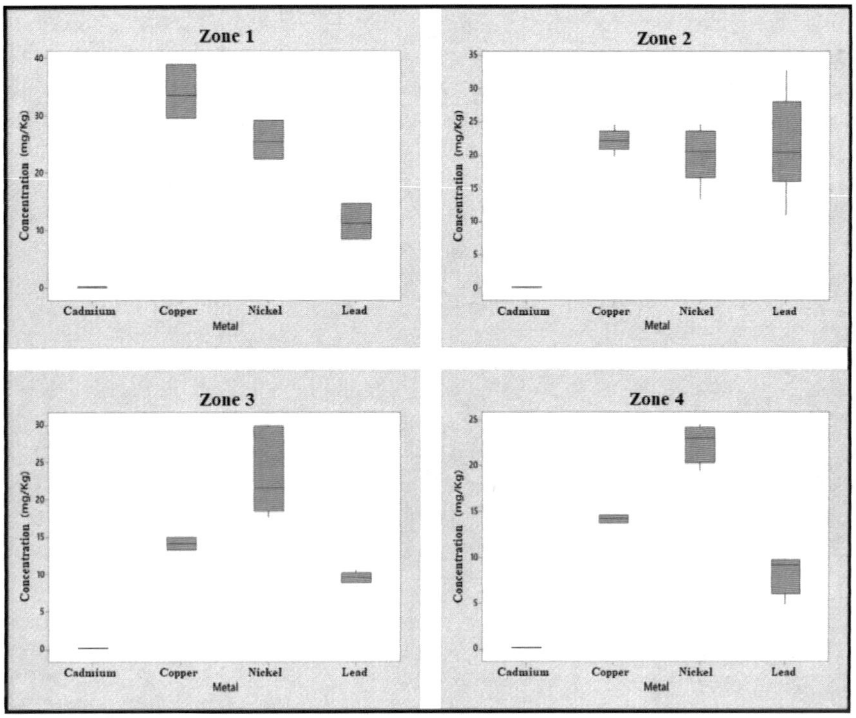

Figure 2. Average concentration (mg/Kg) of copper, nickel, lead and cadmium in four study areas in Latacunga, Ecuador. Zone 1 refers to La Cocha sector. Zone 2 refers to Ignacio Flores Nautical Park and the banks of the Illuchi River. Zone 3 refers to the banks of Cutuchi River close to a food industry and an aluminum fountry. Zone 4 refers to the banks of Saquimala River close to a textile industry, in a rural area.

Table 1. Heavy metal concentration values and permissible levels in the city of Latacunga

Metal	Zone	Permissible reference (mg/Kg)	Concentration (mg/Kg)			Standard deviation
			Minimum	Maximum	Average	
Cu	1	30	29.6	39.0	34.07	4.72
Ni	1	20	22.4	29.3	25.70	3.46
Pb	1	25	8.5	14.6	11.43	3.06
Cu	2	30	19.9	24.6	22.22	1.64
Ni	2	20	13.4	24.6	20.00	4.05
Pb	2	25	11.0	32.6	21.45	7.42
Cu	3	30	13.2	15.0	14.10	0.76
Ni	3	20	17.7	30.0	23.25	5.38
Pb	3	25	8.8	10.6	9.60	0.70
Cu	4	30	13.7	14.7	14.20	0.48
Ni	4	20	19.4	24.5	22.48	2.17
Pb	4	25	4.9	9.8	8.28	2.27

3.2. Spatial distribution of copper, nickel and lead

Figure 3 shows the spatial distribution of each metal analyzed in the 4 study areas. Zone l has the highest copper and nickel concentration values compared to the other studied areas. Zone 2 has a high copper concentration with respect to Zone 3 and Zone 4 and higher lead concentration values with respect to the other study zones. All the zones have similar nickel concentration values. This way, the concentration of copper by zones decreases in order of Zone 1 > Zone 2 > Zone 4 > Zone 3, the concentration of nickel decreases in order of Zone 1 > Zone 3 > Zone 4 > Zone 2 and the concentration of lead falls in order of Zone 2 > Zone 1 > Zone 3 > Zone 4.

After analysis of variance applied to the concentration values of copper, nickel and lead for each of the study areas, the following results were obtained: an F value of 78.44 was obtained for the copper concentration with a p-value of 0.000 (95% confidence level); therefore it is concluded that the copper concentration values are different in each of the sampled areas. For nickel, an F value of 1.35 with a p value of 0.296 (95% confidence level) were obtained, therefore, there is no significant difference between the concentrations of nickel in the sampled areas. Finally, the analysis applied to lead concentration values resulted in an F value of 9.49 and a p value of 0.001 (95% confidence level) resulting in a significant difference between the lead concentrations in the different sampled areas.

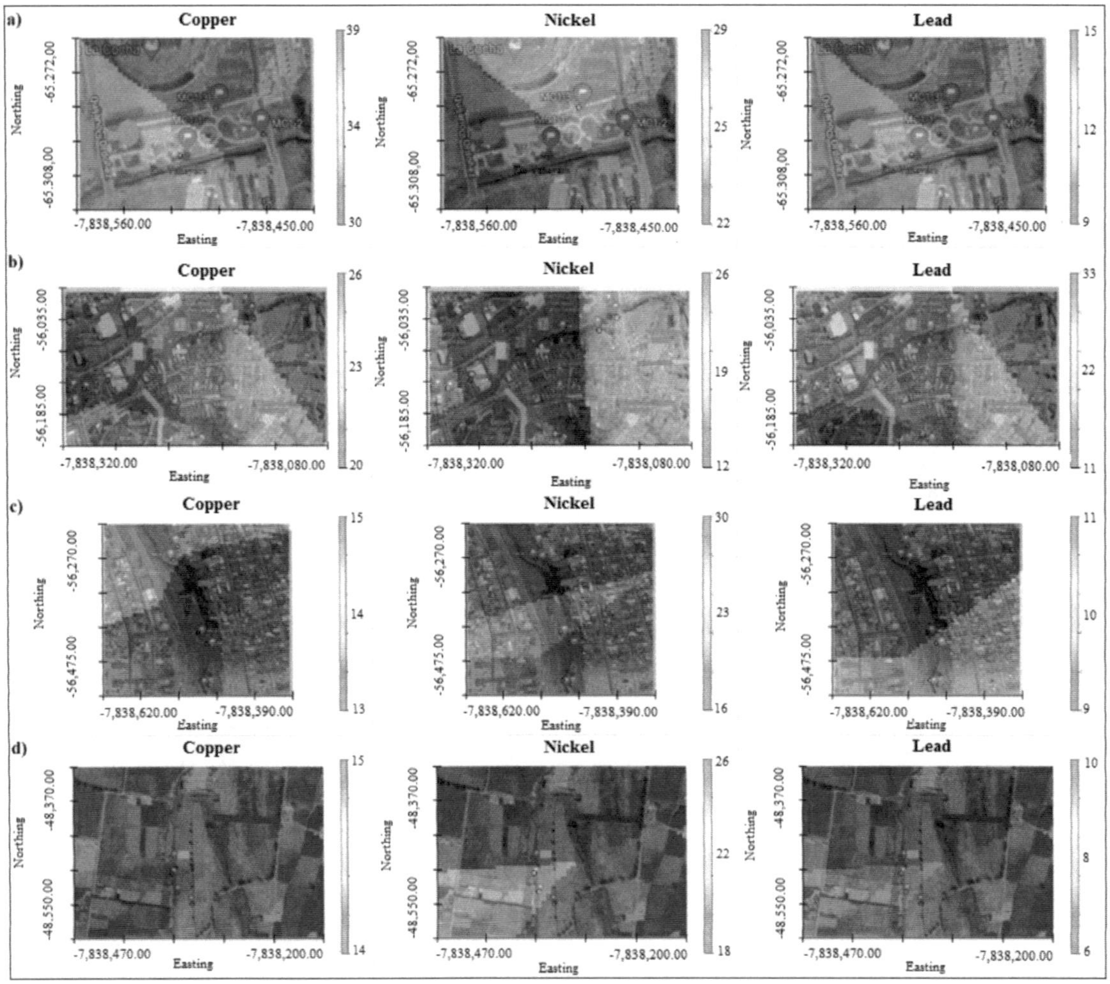

Figure 3. Spatial distribution of copper, nickel and lead in the study area: a) Zone 1, b) Zone 2, c) Zone 3, d) Zone 4. Basemap data © Google 2019.

3.3. Analysis of the concentration of copper, nickel and lead for two particle sizes

The analysis of the concentration of copper, nickel and lead for different particle sizes was performed by analyzing 6 pairs of samples, each pair of samples belonging to the same sample point and the analyzed particle sizes were 38 μm and 1,000 μm as seen in Figure 4.
The analysis of variance of the copper concentration in the different points of soil samples was proposed in order to check if the copper concentration is the same in all the sample points. When obtaining a calculated F value of 13.10 and a p-value of 0.007 (95% confidence level) for the analyzed samples, it is concluded that the copper concentrations of each sample obtained are different from each other. In the paired *t* test applied to the copper concentration data obtained from the analysis of samples of particles of 38 μm and 1,000 μm, a p-value of 0.245 was obtained (confidence level of 95%), therefore it is concluded that there is no significant difference between the copper concentration data obtained from the same sample point but analyzed with two different particle sizes (38 μm and 1,000 μm).
Regarding the analysis of variance carried out for the nickel concentration in the soil samples, it was concluded that nickel concentration values in each analyzed sample do not have a significant difference between their values, this as a result of a calculated F value of 4.28 and a P value of 0.068 (95% confidence level). The nickel concentration data obtained for the two particle sizes (38 μm and 1,000 μm) were analyzed by applying the paired *t* test, obtaining a p-value of 0.214 (95% confidence level), therefore it is concluded that there is no significant difference between the nickel concentration data obtained from the same sample point but analyzed in two particle sizes (38 μm and 1,000 μm).
Regarding the results obtained as a consequence of the application of the analysis of variance, it was determined that the concentration indices of each sample point are different from each other, because the calculated F value of 12.29 and a p-value of 0.008 (95% confidence level). The paired *t* test applied to the data of lead concentration obtained from the analysis in samples of particles of 38 μm and 1,000 μm yielded a p-value of 0.029 (95% confidence level), therefore It is concluded that there is a significant difference between the lead concentration data obtained from the same sample point but analyzed with two different particle sizes (38 μm and 1,000 μm).

4. Conclusions

Zone 1 (La Cocha sector - green area with public access) shows higher concentration values to those established in local regulations due to the presence of copper; Zone 1 and Zone 2 (Ignacio Flores Nautical Park and Illuchi river bank - green areas with public access) present higher concentration values than those established by authorities for lead; all the studied areas present higher concentration values for nickel; finally, no studied area presents Cadmium concentration values exceeding permissible references since every reding was below detection limits.
Despite the fact that Zone 2 does not exceed the maximum permissible limits, it is the zone that has the highest concentration values for copper and lead with respect to Zones 3 and 4. The industries taken into account for this research located in Zone 3 and Zone 4 do not represent a determining factor in soil contamination due to the presence of copper and lead.

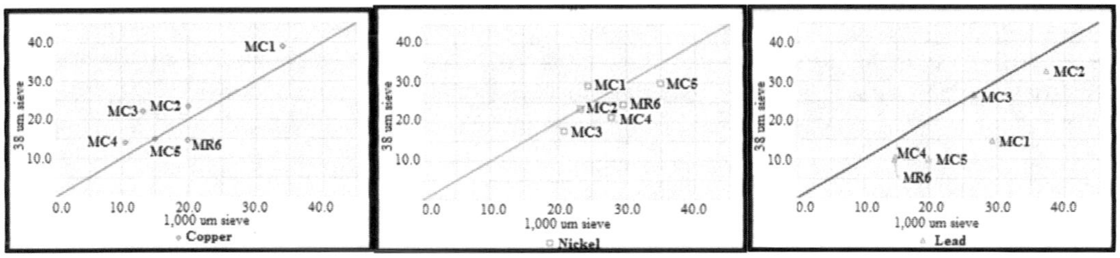

Figure 4. Comparison of the results of the concentration of copper, nickel and lead for the two sizes of particles analyzed

There is a significant difference between the concentrations of the sample points of copper and lead, while nickel does not show a significant difference. Therefore, nickel shows a homogeneous distribution, while copper and lead have a heterogeneous distribution in all study areas.

The data obtained in the comparison of concentrations of heavy metals according to particle size showed that the concentration of copper is greater in the smaller particles (38 µm), while the concentration of nickel and lead is greater in the larger particles (1,000 µm). The paired *t* test showed that there is a significant difference for the lead concentration data, while for the copper and nickel concentration data it was concluded that there is no significant difference with a confidence level of 95% ($\alpha = 0.05$).
Soil contamination due to the presence of nickel in all study areas is related to activities of each industry, since the textile company, the flour processing company and the foundry company use this heavy metal in their production processes. The main affection to human health due to the presence of nickel in the human body is the spread of allergies, in extreme cases it produces stomach damage, adverse effects on the blood and the kidney and in case of extreme inhalation kidney [10].
The main goal of this paper was to identify the nature of nickel, lead, and copper concentrations in soils around Latacunga city. Monitoring soil quality around industries is a job that local authorities must continue and research workers must collaborate with this tasks.

5. References

[1] Y. C. Reyes, I. Vergara, O. E. Torres, M. Díaz, and E. E. González, "Contaminación por metales pesados: implicaciones en salud, ambiente y seguridad alimentaria," *Revista Ingeniería, Investigación y Desarrollo*, vol. 16, no. 2, pp. 66–77, 2016.
[2] C. García, J. L. Moreno, M. T. Hernández, and A. Polo, "Metales pesados y sus implicaciones en la calidad del suelo," *Cienc. y Medio Ambient.*, pp. 125–138, 2002.
[3] C. Zafra, J. Temprano, and I. Tejero, "Concentración y distribución de metales pesados (Pb , Zn , Cu , Cd y Cr) en sedimentos viarios urbanos," pp. 53–62, 2011.
[4] C. Gonzáles, J. Thompson, Y. Martínez, and N. Sánchez, "Concentración de cadmio en partículas de diferentes tamaños de un suelo de la Cuenca del Lago de Valencia," *ISSN 0798-4065*, pp. 73–80, 2010.
[5] P. Sánchez, "Diagnostico Del Plan De Desarrollo Del Cantón Latacunga 2016-2019," 2016.
[6] A. Marín, V. Gonzalez, B. Lapo, E. Molina, and M. Lemus, "Mercury levels in sediments from the coast of El Oro - Ecuador," *Gayana*, vol. 80, no. 2, pp. 147–153, 2016.
[7] USEPA, "Method 3050B. Acid digestion of sediments, sludges, and soils," United States, 1996.
[8] USEPA United States Environmental Protection Agency, "Method 3052 Microwave assisted acid digestion of sileceous and organically based matrices," *Journal of Chemical Information and Modeling*, vol. 53, no. 9. pp. 1689–1699, 2013.
[9] Presidencia de la Republica del Ecuador, *Texto unificado de Legislación Secundaria de Medio Ambiente. Libro VI Anexo 2. Norma de calidad ambiental del recurso suelo y criterios de remediación para suelos contaminados*. 2017, p. 18.
[10] Agencia de Sustancias Tóxicas y Registro de Enfermedades, "Resúmenes de Salud Pública - Níquel (Nickel)," Atlanta, 2014.

Acknowledgments
The authors would like to thank the staff from the laboratory at "Departamento de Metalurgia Extractiva" from Escuela Politécnica Nacional, for their collaboration in the project.

IOP Conf. Series: Earth and Environmental Science **728** (2021) 012007 doi:10.1088/1755-1315/728/1/012007

Natural Radioactivity Evaluation of Local Soil used as Building Materials in Xinchang Section of Beishan Pre-selected Area, Northwest China

Chenhao SHI[1,2], Liang CHEN[2], Yunyan WANG[1*], Liyuan CHAI[1], Guohua QIU[3]

1 School of Metallurgy and Environment, Central South University, Changsha 410083, China
2 Beijing Research Institute of Uranium Geology, Beijing 100029, China
3 CNNC Environment and Protection Co.Ltd., Beijing 100044, China
* E-mail: shichenhao1992@126.com

Abstract. To investigate the radiation background level of jijicao block and rock mass around Xinchang section in the pre-selected area of Beishan, northwest China, the relevant soil samples and data have been obtained, and the natural radioactivity of the soil samples has been measured by gamma-ray spectroscopy with hyper-pure germanium detector. The mean activity concentrations of ^{226}Ra, ^{232}Th and ^{40}K in soil samples are determined as 23.72 (from 11.2 to 48.6 $Bq\cdot kg^{-1}$), 28.72 (from 11.9 to 33.0 $Bq\cdot kg^{-1}$) and 612.93 $Bq\cdot kg^{-1}$ (from 244.0 to 907.0 $Bq\cdot kg^{-1}$), respectively, which are all lower than the Jiuquan background values, among which the concentration of ^{40}K is higher than UNSCEAR 2008 and Beishan background value. The calculated data of radium equivalent, representative level index, external standard, internal standard, and annual effective dose are less than the recommended limits. Therefore, it can be concluded that the building being constructed of the materials is safe for the inhabitants. The findings from this research will be useful to assess the radiation hazards of building materials in humans. It is valuable for the environmental impact assessment of the underground research laboratory of high-level radioactive waste and the assessment of the hazard of radioactive building materials to human body.

1. Introduction

Soil is the basic resource for human survival and plays an extremely important role in the ecological environment. Generally speaking, the soil environment which is not affected by human activities in the natural state will not change after being placed in the humid environment for a certain period of time. Soil environmental conditions will show relatively stable characteristics. The background level of the soil environment is mainly determined and affected by the concentration of natural radionuclides such as rock, ^{226}Ra, ^{232}Th and ^{40}K in the earth's crust [1]. Diversity in geological conditions, bedrock composition, soil development types and topographic distribution in varied areas will lead to significant differences in the concentration of natural radionuclides ^{232}Th, ^{226}Ra and ^{40}K in the soil [2]. However, with the development of the industrial revolution, radionuclides caused by human activities enter the soil environment, which leads to soil radionuclide contamination and long-term radiation hazard to human health [3].

Beishan pre-selected area is characterized by convenient transportation, weak economic conditions (such as lack of farmland, mineral resources, animal, and plant resources) and sparse population in Gansu province, northwest China. It has been selected as the most suitable area for HLW repository in China because of its superior social, economic and natural conditions [4]. In July 2011, the China Atomic Energy Authority, together with the Ministry of Ecology and Environment, approved that the

Beishan area is "the first priority area" for China's HLW repository [5]. In order to use radionuclide soil as the building materials, it is of great significance to investigate and evaluate the radioactive environment of Beishan area.

In this research, the background levels of radionuclides such as ^{226}Ra, ^{232}Th and ^{40}K in the jijicao block and the rock mass around Xinchang section in the pre-selected area of Beishan have been investigated. The radiation hazard of building materials to human body have been evaluated by calculating radium equivalent, gamma index, external standard, internal standard, absorbed dose and annual effective dose, which will be the basis for the construction, operation and management of URL of HLW.

2.Material and methods

2.1. Study area

The pre-selected area of the HLW disposal repository locates in the arid desert region of Gansu Province, northwest China. The study area lies in Xinchang section of Beishan pre-selection area, as shown in Figure 1. It has the characteristics of sparse rainfall, large evaporation, cold and dry wind, low temperature, short summer and long winter. The elevation is 1,600-1,800 m. The relative height difference of the terrain is within 100 m. The soil type on the hillside is mainly aeolian sandy soil, and the soil layer is relatively thin. Precipitation is mostly concentrated in summer, with an average annual precipitation of 85.2 mm and an average annual evaporation of 3,072.9 mm.

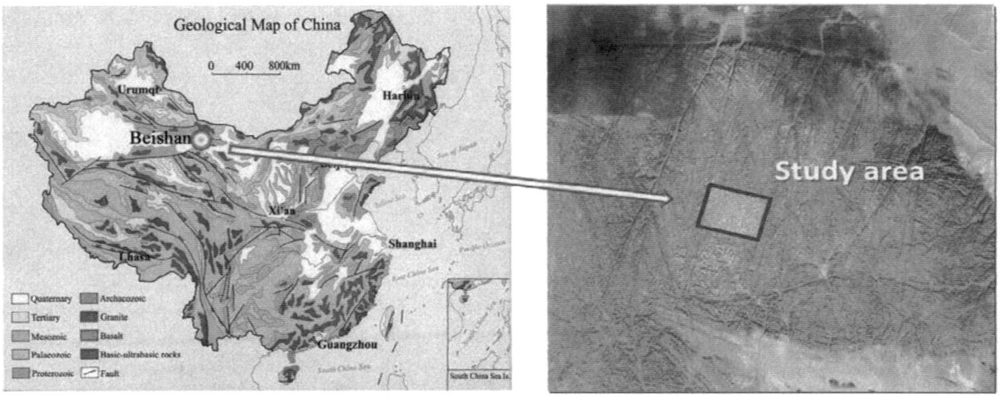

Figure 1. The location of the study area in Beishan, Gansu province [6].

2.2. Sampling and analysis

60 sampling points (No. 108-167) were selected around the jijicao block and the rock mass in the open uncultivated area in spring (Figure 2). The geographic location of the sampling point is recorded using the global positioning system (GPS) [7]. 60 surface soil samples were collected by the five-point sampling method. Stones, roots and other sundries are removed from all soil samples and each sample with the weight no less than 1 kg was put into plastic bag [8]. After the air is discharged, it is sealed and stored in a polythene bag with sampling information indicated, and then sent back to the laboratory for refrigeration.

The samples were weathered in a dry environment for 30 days and dried in an oven at 110 ℃ for 6 hours [9]. The activity concentrations of ^{226}Ra, ^{232}Th and ^{40}K for all homogenized and equilibrium samples were measured by a gamma ray spectrometry (GB/T 11743-2013) by using a high purity Germanium (HPGe) detector (25% relative efficiency) and a coaxial-type vertical dipstick cryostat [10]. The detection limits of ^{226}Ra, ^{232}Th and ^{40}K were 7.05, 0.80, and 0.43 $Bq \cdot kg^{-1}$, respectively [11]. The ^{226}Ra specific activities were estimated from ^{214}Bi (609.3 keV) and ^{214}Pb (295.2 and 352.0 keV), and the ^{232}Th specific activities were estimated from ^{228}Ac (911.1 keV), ^{212}Pb (583.1 keV), and ^{208}Tl (238.6 keV), while the specific activity of ^{40}K was determined directly from its gamma emission at 1460.83 keV. The analytical results of all samples met the quality control requirements of technical criteria for radiation environmental monitoring (HJ/T 61-2001) [12].

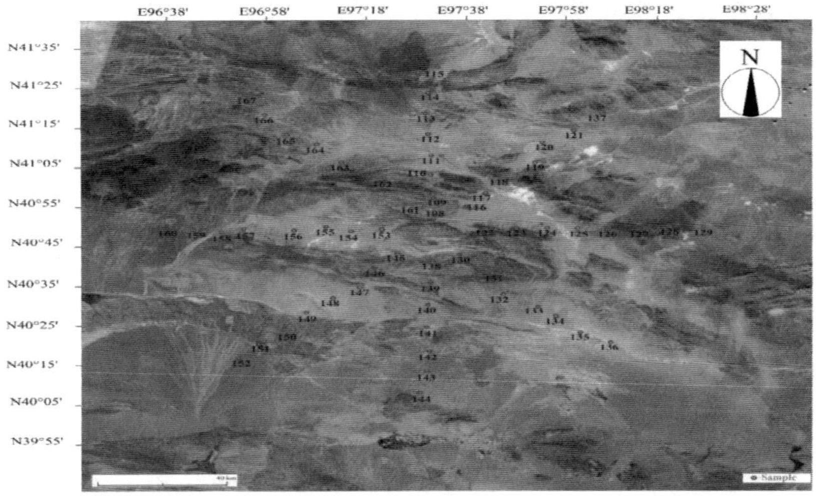

Figure 2. Soil sampling points in the jijicao block and the rock mass around Xinchang area.

2.3. Methodology

2.3.1 Radium equivalent activity (Ra_{eq}). To assess the radiation hazard associated with the building materials, Ra_{eq} was evaluated. Where it is assumed that all decay products of ^{226}Ra and ^{232}Th are in radioactive equilibrium with their precursors. Ra_{eq} is calculated according to the following equation [13]:

$$Ra_{eq} = A_{Ra} + 1.43A_{Th} + 0.077A_{K} \quad (1)$$

where A_{Ra}, A_{Th} and A_{K} are the specific activities of ^{226}Ra, ^{232}Th and ^{40}K in $Bq \cdot kg^{-1}$, respectively. The equation is based on the estimation that 1 $Bq \cdot kg^{-1}$ ^{226}Ra, 0.7 $Bq \cdot kg^{-1}$ ^{232}Th and 13 $Bq \cdot kg^{-1}$ ^{40}K produce the same gamma-ray dose rates [14]. Ra_{eq} is related to both internal doses due to the radon and external gamma doses, and it should have the highest value of 370 $Bq \cdot kg^{-1}$ for safe use in building materials [15].

2.3.2 Representative level index (I_r). Representative level index (I_r) used to estimate the standard of gamma radiation hazard associated with the natural radionuclides in specific building materials may exceed 1 $mSv \cdot y^{-1}$. It is calculated using the following equation [16]:

$$I_r = C_{Ra} / 300 + C_{Th} / 200 + C_K / 3000 \quad (2)$$

where C_{Ra}, C_{Th} and C_K in $Bq \cdot kg^{-1}$ are the concentration of ^{226}Ra, ^{232}Th, and ^{40}K, respectively.

2.3.3 External hazard index (H_{ex}) and internal hazard index (H_{in}). The two hazard indexes (H_{ex} and H_{in}) are used to characterize the external and internal hazards due to the emitted gamma radiation. The primary objectives of H_{ex} and H_{in} are to limit the radiation dose to the equivalent limit of 1 $mSv \cdot y^{-1}$, and the two equations are as follows [17]:

$$H_{ex} = A_{Ra} / 370 + A_{Th} / 259 + A_K / 4810 \quad (3)$$

$$H_{in} = A_{Ra} / 185 + A_{Th} / 259 + A_K / 4810 \quad (4)$$

2.3.4 Absorbed gamma dose rate (D_G) and annual effective dose rate (D_E). The outdoor absorbed dose rate ($nGy \cdot h^{-1}$) in air from terrestrial gamma radiation at 1 m above the ground is calculated after applying the conversion factors (in $nGy \cdot h^{-1}$ per $Bq \cdot kg^{-1}$) to transform A_{Ra}, A_{Th} and A_K into D_G according to the equation provided by UNSCEAR [18]:

$$D_G = 0.08A_K + 0.92A_{Ra} + 1.1A_{Th} \quad (5)$$

To estimate D_E, it is necessary to use the conversion coefficient from the absorbed dose in air to the effective dose (0.7 Sv·Gy^{-1}) and the outdoor occupancy factor (0.2 Sv·Gy^{-1}) proposed in UNSCEAR. The annual time is calculated by 8760 hours. Therefore, D_E is determined as follows [19]:

$$D_E = D_G \times 0.2 \times 0.7 \times 8760 \times 10^{-6} \quad (6)$$

In UNSECEAR 2008, it is stipulated that the standard values of D_G and D_E are 84 nGy·h^{-1} and 0.48 mSv·y^{-1}, respectively [20].

3. Results and discussion

3.1 The concentration of natural radionuclides in building materials

Table 1. ^{226}Ra, ^{232}Th, ^{40}K activity concentration (Bq·kg^{-1}) of the soil samples in Xinchang section around the jijicao block and the rock mass.

Building material	Sample number	Sample size	^{226}Ra Mean	^{226}Ra Range From	to	^{232}Th Mean	^{232}Th Range From	to	^{40}K Mean	^{40}K Range From	to
Sand stone	108,109,110,112,113,114,115, 116,118, 119,120,122,123,124,126,128, 130,132,134, 135,138,139,143,144,145,146, 147,149,153, 156,157,158,159,160,167	35	24.15	12.5	32.8	28.85	14.0	48.6	611.71	291.0	907.0
Sand soil	121,127,129,137,140,148,150, 151,152 154,155,162,163,164,165,166	16	23.47	11.9	33.0	27.68	11.2	47.2	623.44	244.0	861.0
Stiff soil	111,117,125	3	20.0	12.2	26.6	28.93	19.9	39.0	682.0	648.0	725.0
Mollisol	131,136,161	3	22.53	21.0	23.4	28.6	23.0	34.0	524.33	462.0	615.0
White soil	133,141,142	3	25.0	21.4	31.6	32.67	21.9	40.1	590.67	418.0	763.0
Total	—	60	23.72	11.9	33.0	28.72	11.2	48.6	612.93	244.0	907.0
Beishan background value [21]	—	—	34.3	—	—	38.8	—	—	652.1	—	—
Jiuquan background value [22]	—	—	25.4	—	—	35.7	—	—	429.2	—	—
UNSCEAR 2008 [23]	—	—	50.0	—	—	50.0	—	—	500.0	—	—

Table 1 lists the ranges and the mean values of the activity concentrations of ^{226}Ra, ^{232}Th and ^{40}K for sixty types of building materials in Xinchang section around the jijicao block and the rock mass. From Table 1, the mean value of radionuclides of ^{226}Ra, ^{232}Th and ^{40}K are 23.72, 28.72 and 612.93 Bq·kg^{-1} in soil samples, respectively. The concentration of ^{40}K in soil sample is higher than that prescribed in UNSCEAR 2008 and Beishan background value, but is lower than Jiuquan background value. The mean value of ^{226}Ra, ^{232}Th and ^{40}K in sand stone, sand soil, stiff soil, mollisol and white soil used as building materials are 24.15, 28.85, 611.71 Bq·kg^{-1}, 23.47, 27.68, 623.44 Bq·kg^{-1}, 20.0, 28.93, 682.0 Bq·kg^{-1}, 22.53, 28.6, 524.33 Bq·kg^{-1} and 25.0, 32.67, 590.67 Bq·kg^{-1}, respectively. According to the mean value of ^{226}Ra, the soil active concentration of ^{226}Ra is used as building material decrease following the order of white soil>sand stone>sand soil>mollisol>stiff soil. The mean radionuclide concentrations of ^{232}Th is showed as white soil>stiff soil>sand stone>mollisol>sand soil. And the mean value of ^{40}K increases in order of mollisol<white soil<sand stone<sand soil<stiff soil.

Table 1 also shows that the mean radionuclide concentrations of ^{226}Ra and ^{232}Th of all soils used as building materials are lower than the background values of Beishan, Jiuquan and the value specified in UNSCEAR 2008, while the concentration of ^{40}K of stiff soil is slightly higher than Beishan background value. ^{40}K concentrations of sand stone, sand soil, stiff soil, mollisol and white soil are all higher than the value specified in UNSCEAR 2008, and there are 1.3 to 1.45 times higher than Jiuquan background value.

Table 2 shows the comparison of the radioactivity concentration of different types of soil in China and other parts of the world. The ranges of the mean values of the natural radionuclide concentration in building materials differ from one country to another, depending on the soil used for their formation. Except for ^{40}K, the concentration of radionuclides of ^{226}Ra and ^{232}Th in the soil types used as building materials in Xinchang section around the jijicao block and the rock mass is lower than the mean value in China. ^{226}Ra concentration of the different types of soil from Xinchang section is higher than the mean value in Pakistan, Cuba and Greece, but is lower than that in Iran and Brazil. The concentration of ^{232}Th is higher than that in Cuba, Greece, and Iraq, however it is lower than that in Pakistan and Brazil. The radionuclide concentration of ^{40}K is 1.4 to 4.5 times higher than those in the countries of Brazil, Pakistan, Greece, Cuba and Iraq. The previous research results revealed that the concentration of ^{40}K exceeds the Jiuquan background value and UNSCEAR 2008 may be caused by nuclear research and radioactive waste disposal activities in nearby area [24].

3.2 Radium equivalent activity (Ra_{eq}) in building materials

The mean value of the radium equivalent activity (Ra_{eq}) for sample soils of building materials is presented in Figure 3. The maximum Ra_{eq} is 117.2 Bq·kg^{-1} in white soil, while the minimum Ra_{eq} found in mollisol is approximately 103.805 Bq·kg^{-1}. The mean value of Ra_{eq} are 111.05, 112.51, 113.89 Bq·kg^{-1} in sand soil, sand stone and stiff soil, respectively. The estimated mean value of Ra_{eq} in the soil samples is 111.99 Bq·kg^{-1}, which is significantly less than the upper limit of 370 Bq·kg^{-1}, and it does not pose any radiological hazard when used for the construction of buildings.

Table 2. Comparison of radioactivity concentration of soil types in China and other parts of the world.

Country	Mean value of activity concentration (Bq·kg^{-1})			References
	^{226}Ra	^{232}Th	^{40}K	
China	32	41	440	[25]
Pakistan	20	29	383	[26]
Cuba	17	16	208	[27]
Greece	18	17	367	[28]
Iraq	43.57	1.98	135.02	[29]
Brazil	36.3	49.4	444.32	[30]
Worldwide	32	45	420	[31]

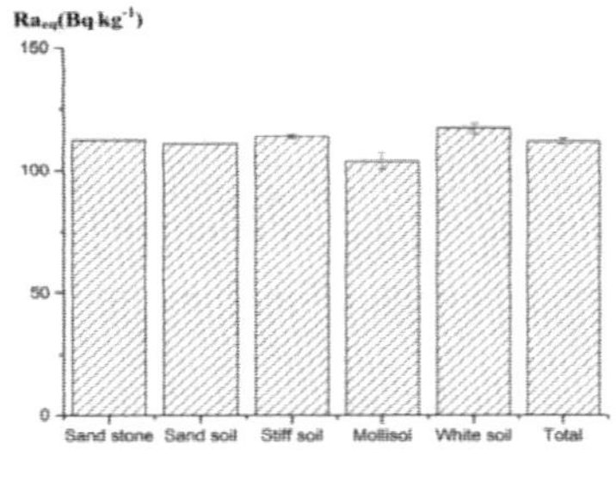

Figure 3. Mean values of Ra_{eq} of building materials.

3.3 Representative level index (I_r) in building materials

The mean value of the activity utilization index for different types of soil of building materials in Xinchang section around the jijicao block and the rock mass is shown in Figure 4. The mean activity utilization of I_r are 0.43, 0.42, 0.44, 0.39 and 0.44 mSv • y^{-1} in sand stone, sand soil, stiff soil, mollisol and white soil, respectively. And the mean value of I_r of different types of soil used in building material is 0.43. I_r <1 implies that the building materials in Xinchang section around the jijicao block and the rock mass are less than the specified value of 1 mSv·y^{-1}. The result indicates that the five types of soil are safe for using as building materials.

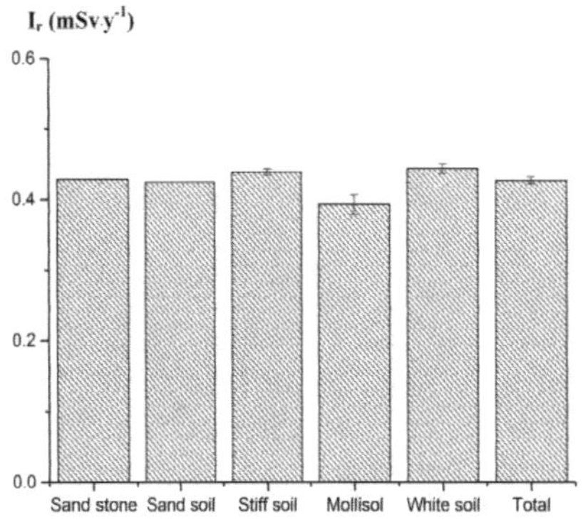

Figure 4. Mean values of I_r of building materials.

3.4 External hazard index (H_{ex}) and internal hazard index (H_{in}) in building materials

The calculated mean values of H_{ex} and H_{in} for all types of building materials are presented in Figure 5. The higher H_{ex} of 0.32 and 0.31 are found in white soil and stiff soil, respectively. The obtained mean values of sand stone and mollisol used as building materials in this study are lower than the total mean value. The mean value of H_{ex} estimated in total is 0.3 mSv·y^{-1}, which is significantly less than the upper limit of 1 mSv·y^{-1}. While the mean values of H_{in} is 0.37 mSv·y^{-1}, and the mean value of H_{in} are 0.37, 0.36, 0.36, 0.34 and 0.38 mSv·y^{-1} in sand stone, sand soil, stiff soil, mollisol and white soil, respectively, which are all less than the recommended level of 1 mSv·y^{-1}. We can say that the radiation hazard is insignificant for the population. Therefore, the five types of soil in this study can be safely used in the construction of buildings.

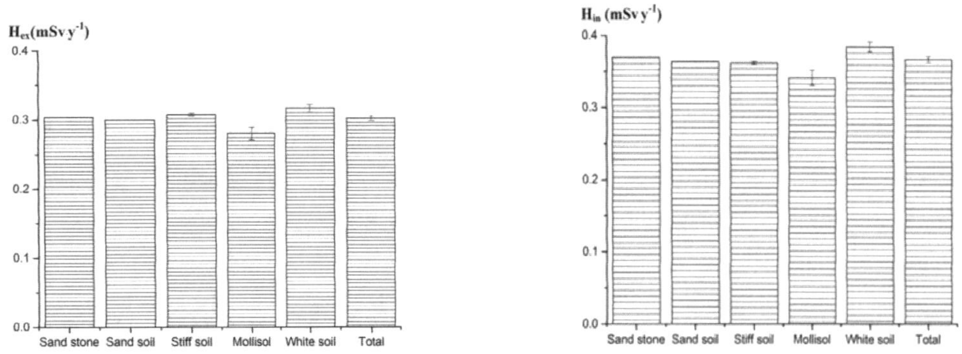

Figure 5. Mean values of H_{ex} and H_{in} of building materials.

3.5 Absorbed gamma dose rate (D_G) and annual effective dose rate (D_E) in building materials

The mean values of D_G in air for different types of soil used for building materials are shown in Figure 6. The maximum D_G is 106.19 nGy·h^{-1} in white soil, while the minimum D_G found in mollisol is approximately 94.14 nGy·h^{-1}. The estimated mean value of D_G in the soil samples is 102.45 nGy·h^{-1}, which is slightly higher than world (populated-weighted) (UNSCEAR 2008) indoor D_G of 84 nGy·h^{-1} by about 1.2 times. The mean annual outdoor D_E of different types of soil for building material are also given in Figure 6. These values vary from 0.12 mSv·y^{-1} for mollisol to 0.13 mSv·y^{-1} for white soil. The estimated mean value of the annual D_E of 0.13 mSv·y^{-1} is less than the permissible limit of 0.48 mSv·y^{-1}. Figure 6 also presents the mean D_G and D_E of 102.89, 101.91, 104.79 nGy·h^{-1} and 0.13, 0.12, 0.13 mSv·y^{-1} in sand soil, sand stone and stiff soil, respectively. The results indicate that the various types of building materials along with the D_G and the D_E, respectively. According to the current research progress, the five types of soil used as building materials in Xinchang section around the jijicao block and the rock mass shall be classified and used according to limits of radionuclides in building materials (GB6566-2010)[32].

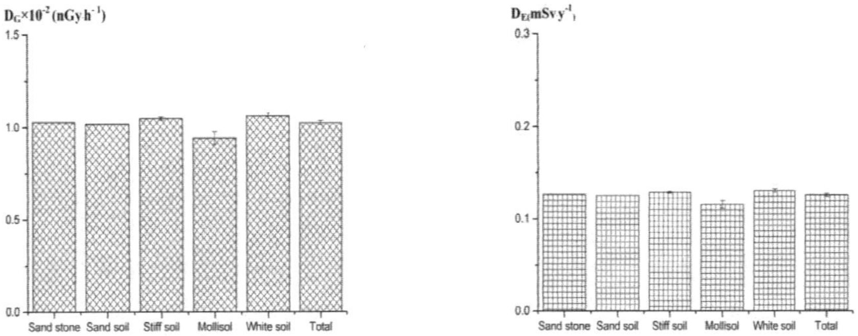

Figure 6. Mean values of D_G and D_E of building materials.

At present, there are few studies on the effects of the soil types of radiation hazard indexes used as building materials. We compared the effects of the radiation hazard indexes related to the desert areas on the granite used as building materials. Table 3 shows the comparison of the radiation hazards of the granite samples in China and other parts of the world. The mean value of Ra_{eq}, I_r, H_{ex}, H_{in}, D_G and D_E in the soil types used as building materials in Xinchang section around the jijicao block and the rock mass is lower than the mean value of the granite samples used as building materials in China, Pakistan and Saudi Arabia, but it is higher than that in Iran, Italy and Egypt. The previous research results revealed that granite is generally considered to be a building material with high concentration of radionuclides, but its radioactive background value is still related to its type [33].

Table 3. Comparison of Ra_{eq}, I_r, H_{ex}, H_{in}, D_G and D_E of the granite samples in China and other parts of the world [34].

Country	Radiation hazards					
	Ra_{eq} (Bq·kg^{-1})	I_r (mSv·y^{-1})	H_{ex} (mSv·y^{-1})	H_{in} (mSv·y^{-1})	D_G (nGy·h^{-1})	D_E (mSv·y^{-1})
China	228.35	0.863	0.617	0.741	109.78	0.539
Pakistan	273.52	1.027	0.739	0.914	130.64	0.641
Iran	4.63	0.017	0.012	0.017	2.19	0.011
Italy	14.03	0.049	0.038	0.048	6.28	0.031
Saudi Arabia	271.61	1.035	0.733	0.483	131.75	0.646
Egypt	7.31	0.027	0.019	0.028	3.47	0.017
Worldwide	129.84	0.492	0.351	0.914	62.60	0.307

4. Conclusions

The radioactivity concentrations of ^{226}Ra, ^{232}Th and ^{40}K in Xinchang section around the jijicao block and the rock mass are all lower than the background values of Jiuquan, among which the concentration of ^{40}K is higher than the background values of Beishan and the specified value in UNSCEAR 2008. The soils are divided into five types, in which the concentrations of ^{40}K radionuclides of sand stone, sand soil, stiff soil, mollisol and white soil are all higher than the background values of Jiuquan and the specified value in UNSEAR 2008, and the mean values of stiff soil are also higher than the background values of Beishan. The excessive concentration of ^{40}K may be related to nuclear research and radioactive waste disposal activities in nearby areas.

The mean values of Ra_{eq} and I_r for all types of building materials are significantly lower than the upper limit of 370 Bq·kg^{-1} and 1 mSv·y^{-1}, respectively. H_{ex} and H_{in} of all kinds of soils used as building materials are less than the upper limit of the recommended level of 1 mSv·y^{-1}, which indicated that the radiation hazard had few effect on the residence and activities of the population. D_E of the soil (0.48 mSv·y^{-1}) is less than the upper limit of 1 mSv·y^{-1}, and the D_G of five types of soil used as building materials are slightly higher than world mean value (populated-weighted) (UNSCEAR 2008) indoor D_G of 84 nGy·h^{-1}. We suggest that the five types of soil in Xinchang section around the jijicao block and the rock mass shall be classified and used according to limits of radionuclides in building materials.

The works of this research can provide reference for the construction of a URL and the environmental impact assessment of HLW, and also provide a quantitative basis for the assessment of human health hazards caused by radioactive building materials.

5. References

[1] FARIDEH A B, Farid M, REZA F, et al. Distribution of natural radionuclides and assessment of the associated radiological hazards in the rock and soil samples from a high-level natural radiation area, Northern Iran[J]. Journal of Radioanalytical and Nuclear Chemistry, 2019, 332, 2091-2103.

[2] PARGIN B, ROHIT M, RAJAN J, et al. Estimation of ^{222}Rn exhalation rate and assessment of radiological risk from activity concentration of ^{226}Ra, ^{232}Th and ^{40}K[J]. Journal of Geochemical Exploration, Part B, 2018, 184, 304-310.

[3] BOGUSLAAW M. NORM contaminated area identification using radionuclides activity concentration pattern in a soil profile[J]. Journal of Environmental Radioactivity, 2017, 173, 102-111.

[4] LIU J F, CHEN L, WANG C P, et al. Characterizing the mechanical tensile behavior of Beishan granite with different experimental methods[J]. International Journal of Rock Mechanics and Mining Sciences, 2014, 69, 50-58.

[5] WANG J. On area-specific underground research laboratory for geological disposal of high-level radioactive waste in China[J]. Journal of Rock Mechanics and Geotechnical Engineering, 2014, 6(2), 99-104.

[6] WANG J, CHEN L, SU R, et al. The Beishan underground research laboratory for geological disposal of high-level radioactive waste in China: Planning, site selection, site characterization and in situ tests[J]. Journal of Rock Mechanics and Geotechnical Engineering, 2018, 10(3), 411-435.

[7] OLUYIDE S O, TCHOKOSSA P, OROSUN M M, et al. Natural radioactivity and radiological impact assessment of soil, food and water around iron and steel smelting area in Fashina village, Ile-Ife, Osun State, Nigeria[J]. Journal of Applied Sciences and Environmental Management, 2019, 23(1), 135-143.

[8] SERPIL A, FATIH K, YELIZ N. A suggestion to radiological hazards characterization of ^{226}Ra, ^{232}Th, ^{40}K and ^{137}Cs: spatial distribution modelling[J]. Journal of Hazardous Materials, 2019, 353, 476-489.

[9] RITSANANUWAT R, CHANYOTHA S, KRANROD C, et al. Transfer factor of ^{226}Ra, ^{232}Th and ^{40}K from soil to Alpinia Galangal plant grown in northern Thailand[J]. Journal of Physics: Conference Series, 2017, 860(1), 1-9.

[10] HARIBALA, HU B, WANG C G, et al. Assessment of radioactive materials and heavy metals in the surface soil around uranium mining area of Tongliao, China[J]. Ecotoxicology and Environmental Safety, 2016, 130, 185-192.
[11] WANG J , LIU J , Chen Y , et al. Technologically elevated natural radioactivity and assessment of dose to workers around a granitic uranium deposit area, China[J]. Journal of Radioanalytical & Nuclear Chemistry, 2016, 310(2), 733-741.
[12] YANG Q, WANG L, DU H Y, et al. Investigation of radioactive level of drinking water sources in the Upper Yangtze River of Chongqing city[J]. Journal of Radioanalytical and Nuclear Chemistry, 2019, 321, 141-149.
[13] MALANCA A, PESSINA V, DALLARA G. Radionuclide content of building materials and gamma Ray dose rates in dwellings of Rio Grande Do Norte, Brazil[J]. Radiation Protection Dosimetry, 1993, 48(2), 199-203.
[14] Lu X, Chao S, Yang F. Determination of natural radioactivity and associated radiation hazard in building materials used in Weinan, China[J]. Radiation Physics & Chemistry, 2014, 99(4): 62–67.
[15] EI-TAHER A. Gamma spectroscopic analysis and associated radiation hazards of building materials used in Egypt[J]. Radiation Protection Dosimetry, 2010, 138(2), 166-173.
[16] AL-MUGREN K S, EL-TAHER A. Risk assessment of some radioactive and elemental content from cement and phosphate fertilizer consumer in Saudi Arabia[J]. Journal of Environmental Science and Technology, 2016, 9(4), 323-328.
[17] BEREKA J, MATHEW P J. Natural radioactivity of Australian building materials, industrial wastes and by-products[J]. Health Physics, 1985, 48(1), 87-95.
[18] KEHINDE A, CHRISTOPHER O, BOSEDE B O. Measurement of natural radioactivity and radiological hazard evaluation in the soil samples collected from Owo, Ondo State, Nigeria[J]. Journal of Radiation Research and Applied Sciences, 2019, 12(1), 200-209.
[19] MAHUR A K, KUMAR R, SENGUPTA D, et al. Estimation of radon exhalation rate, natural radioactivity and radiation doses in fly ash samples from Durgapur thermal power plant, West Bengal, India[J]. Journal of Environmental Radioactivity, 2008, 99(8), 1289-1293.
[20] UNSCEAR. Effects of atomic radiation sources and risks of Ionizing radiation, report to the general assembly with annexes[M]. United National Scientific Committee, United Nations, New York, 2000.
[21] QIU G H, LI Q, JIN Y X. Investigation on present environmental protection and land use of Jijicao and Xinchang rock mass in Beishan region, Gansu Province[J]. World Nuclear Geoscience, 2014, 31(1), 178-183. (in Chinese)
[22] REN X Y, BAI S M, SUN Q, et al. Investigation of natural radionuclide contents in soil in Gansu Province[J]. Radiation Protection, 1995, 15(5), 381-384. (in Chinese)
[23] UNSCEAR. Report of the United Nations Scientific Committee on the effects of atomic radiation: Fifty-sixth Session[M]. United Nations Publications, United National Scientific Committee, United Nations, New York, 2008.
[24] WEI X, GUO Y, CHEN H L, et al. Rock Mass characteristics in Beishan, a preselected area for China's high-level radioactive waste disposal[J]. Acta Geologica Sinica, 2019, 93(2), 362-372.
[25] CHAO S, LU X, ZHANG M, et al. Natural radioactivity level and radiological hazard assessment of commonly used building material in Xining, China[J]. Journal of Radioanalytical and Nuclear Chemistry, 2014, 300(2), 879-885.
[26] FAHEEM M, MATIULLAH S A. Assessment of radiological hazards due to the natural radioactivity in soil and building material samples collected from six districts of the Punjab province-Pakistan[J]. Radiation Measurements, 2008, 43(8), 1443-1447.
[27] FLORES O B, ESTRADA A M, SUAREZ R R, et al. Natural radionuclide content in building materials and gamma dose rate in dwellings in Cuba[J]. Journal of Environmental Radioactivity, 2008, 99(12), 1834-1837.
[28] STOULOS S, MANOLOPOULOU M, PAPASTEFANOU C. Assessment of natural radiation exposure and radon exhalation from building materials in Greece[J]. Journal of Environmental Radioactivity, 2003, 69(3), 225-240.

[29] HUSSAIN H H, HUSSAIN R O, YOUSEF R M, et al. Natural radioactivity of some local building materials in the middle Euphrates of Iraq[J]. Journal of Radioanalytical & Nuclear Chemistry, 2010, 284(1), 43-47.
[30] VEIGA R, SANCHES N, ANJOS R M, et al. Measurement of natural radioactivity in Brazilian beach sands[J]. Radiation Measurements, 2006, 41(2), 189-196.
[31] LU X, Zhang X. ^{226}Ra, ^{232}Th and ^{40}K activities in soils of Cuihua Mountain National Geological Park, China[J]. Environmental Geology, 2008, 56(2), 353-357.
[32] WANG X, FENG Q Y, SUN R Y, et al. Radioactivity of natural nuclides (^{40}K, ^{238}U, ^{232}Th, ^{226}Ra) in coals from Eastern Yunnan, China[J]. Minerals, 2015, 5(4), 637-646.
[33] ASHRAFI S , JAHANBAKHSH O. Measurement of natural radioactivity of Iranian granite samples using beta-gamma coincidence spectrometer and maximum likelihood method[J]. Environmental Geology, 2019, 78(15): 437.1-437.8.
[34] AI-ZAHRANI J H. Estimation of natural radioactivity in local and imported polished granite used as building materials in Saudi Arabia[J]. Journal of Radiation Research and Applied Sciences, 2017, 10, 241-245.

Acknowledgement
This study is supported by special project on decommission of nuclear facilities and disposal of radioactive wastes (SASTIND No.2 Department No. [2014] 305).

Investigation on the Development of Fuel Consumption for China's Gasoline Passenger Car in Recent Years

Hanzhengnan Yu, Yu Liu* and Jingyuan Li

China Automotive Technology and Research Center Co., Ltd., Tianjin, China
E-mail: liuyu2016@catarc.ac.cn

Abstract. As the application of several energy saving technologies and the continuous tightening fuel consumption standards, the fuel consumption of China's gasoline passenger cars is decreasing continuously. In this paper, the development of the standards related to fuel consumption was reviewed firstly, and the development of the fuel consumption for China's gasoline passenger cars has been investigated based on the declared fuel consumption data of over 4800 vehicle models produced or imported in recent 5 years (2014-2018). Results showed that the average fuel consumption in 2018 decreased 6.4% compared to 2014, about 1.6% per year. In additions, the development of gasoline passenger cars with turbocharge engines and auto transmissions had a more significant effect on the fuel consumption reduction, and the annual decreasing rate reached 2.4% per year and 3.6% per year, respectively.

1. Introduction

China, the world's largest auto market, produced more than 25.72 million automobiles in 2019, which contained 21.36 million passenger cars with a share of 83.0%, and the China's passenger car ownership reached 221 million by the end of 2019 [1-2]. As a result, the contribution of the passenger cars to the nation's transportation fuel consumption is increasing significantly, and caused the problems of air quality and congestion in many large cities. Facing the increasing pressure on the nation's goals of oil independence and climate change mitigation, documents of Energy saving and New Energy Vehicles Industry Strategic Planning (2012-2020) has been published in 2012. According to the documents, the average fuel consumption of the new passenger car should obtain 5.0 L/100km in 2020 and 4.0L/100km in 2025 [3]. In order to achieve these targets, the per-vehicle fuel consumption limits (GB 19578) and the corporate average fuel consumption (CAFC) targets (GB 27999) are strengthening simultaneously [4-5]. In recent years, the fuel consumption of China's gasoline passenger cars decreased continuously by using the advanced technologies including engine and transmission technologies [6-7]. In this paper, the development of fuel consumption of China's gasoline passenger cars in recent years has been investigated. The development of the related standards was introduced firstly, and the development trend of the fuel consumption has been quantified based on the declared fuel consumption data of over 4800 gasoline passenger car models produced or imported in recent 5 years (2014-2018). Finally, the effects of typical energy saving technologies including turbocharge and advanced transmission on the fuel consumption reduction have also been discussed.

2. China's passenger car fuel consumption standards in various phases

China's fuel consumption standards for passenger car was first adopted in 2004, which established Phase I and Phase II per-vehicle fuel consumption limits. Since that, the standards went through five phases including the update of the per-vehicle fuel consumption limits and CAFC targets, as shown in Figure 1.

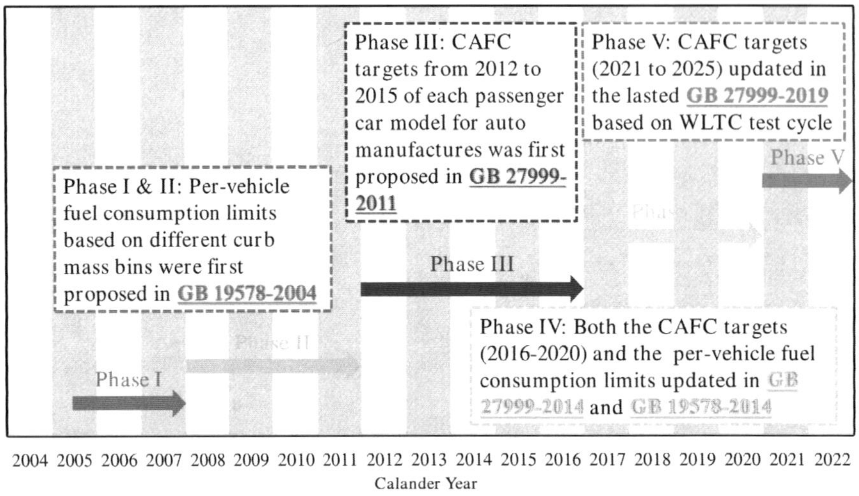

Figure 1. Development of China's passenger car fuel consumption standards

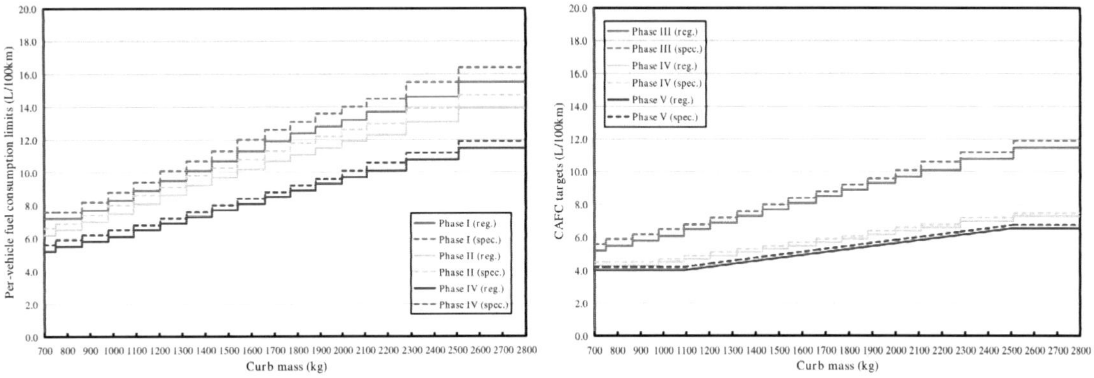

Figure 2. Variations of per-vehicle fuel consumption limits and CAFC targets in various phases

In Phases I (2005-2007) and Phase II (2008-2011), the per-vehicle fuel consumption limits were divided into regular vehicles and special featured vehicles (vehicles with three or more rows of seating and vehicles with automatic transmissions), and both were set based on different curb mass bins.

In Phase III (2012-2015), the CAFC targets from 2012 to 2015 of each passenger car model for auto manufactures were first introduced, the set of target value was based on the same curb mass bin of Phase I and II, and the passenger cars were also divided into regular vehicles and special featured vehicles same as Phase I and II. The addition of CAFC target was supposed to further decrease the fleet-average fuel consumption.

In Phase IV (2016-2020), both the standards of per-vehicle fuel consumption limits and CAFC targets were updated. The per-vehicle fuel consumption limits were modified according to the CAFC targets in Phase III, and the CAFC targets from 2016 to 2020 were further strengthened. In addition, the preferential policy to vehicles with automatic transmissions were eliminated, the special featured vehicles was only defined as vehicles with three or more rows of seating.

In Phase V (2021-), the CAFC targets from 2021 to 2025 and after were strengthened continuously according to the lasted standard of GB 27999-2019, and it worth noting that, instead of the step curve used to calculate CAFC target of each vehicle model in the former standards, linear curve related to the curb mass was used in order to evaluate the fuel consumption more accurate. Besides, the standard related to the per-vehicle fuel consumption limits (GB 19578) would be published officially in the first half of 2020. Except for the variations of per-vehicle fuel consumption limits and CAFC targets, the fuel consumption measurement methods had also been updated, the test cycle was varied from NEDC (New European Driving Cycle) to WLTC (Worldwide Harmonized Light Vehicles Test Cycle). The variations of per-vehicle fuel consumption limits and CAFC targets of these five phases are shown in Figure 2.

3. Results and discussion

In this part, the declared fuel consumption data of over 4800 gasoline passenger car models produced or imported in recent 5 years (2014-2018) has been analyzed. The database includes vehicle's information such as vehicle mass, gasoline engine, transmission and fuel consumption as shown in Table 1. Based on these data, the development trends of the fuel consumption for China's gasoline passenger car in recent years have been investigated first, and the effects of typical energy saving technologies including turbocharge and advanced transmission on the fuel consumption reduction have also been discussed.

Table 1. Information of vehicle in database

Information type	Details
Vehicle mass (kg)	Curb mass/Maximum design mass
Gasoline engine	Engine type (Naturally aspirated/ turbocharge) Displacement (L) Rated power (kW)
Transmission	Transmission type (Manual transmission MT/Auto transmission AT/ Continuously Variable Transmission CVT/ Dual Clutch Transmission DCT)
Fuel consumption	Unban, suburb and combined cycles under NEDC

3.1. Development trends of the fuel consumption for gasoline passenger car

As we all known that NEDC contains an urban cycle and a suburb cycle used to simulate the real-road driving. The declare fuel consumption of a gasoline passenger car in china includes the fuel consumption of urban cycle, suburb cycle and the combination of both. The variations of these three fuel consumption average values from 2014 to 2018 have been calculated as shown in Figure 3. Results showed that the average fuel consumption of NEDC in 2018 was 7.13 L/100km, which was 6.4% lower than 2014, and the annual decreasing rate was about 1.6% per year. It also indicated that the fuel consumption decreasing under the urban cycle was the most obvious whose annual decreasing rate reached 2.0% per year, and the decreasing rate under suburb cycle was smaller, only 1.1% per year.

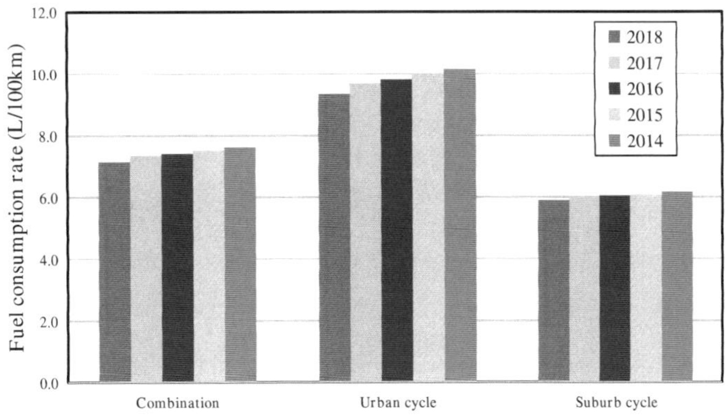

Figure 3. Variations of average fuel consumption under unban cycle, suburb cycle and NEDC

Because that the cub mass of the passenger car directly affected the fuel consumption results, and the standard limits of the fuel consumption were also set based on different curb mass bins, the variations of the average fuel consumption under each curb mass bin have been giving in Figure 4. For either cycle, the fuel consumption increased linearly with the increase of curb mass. The annual decreasing trends of the fuel consumption were also observed, and the gap between the fuel consumption of each year enlarged as the increase of the curb mass especially under urban cycle whose largest gap reached 4.13 L/100km and the annual decreasing rate was about 5.38% per year. The results also indicated that the fuel consumption reduction was more apparent for the gasoline passenger cars with higher curb mass.

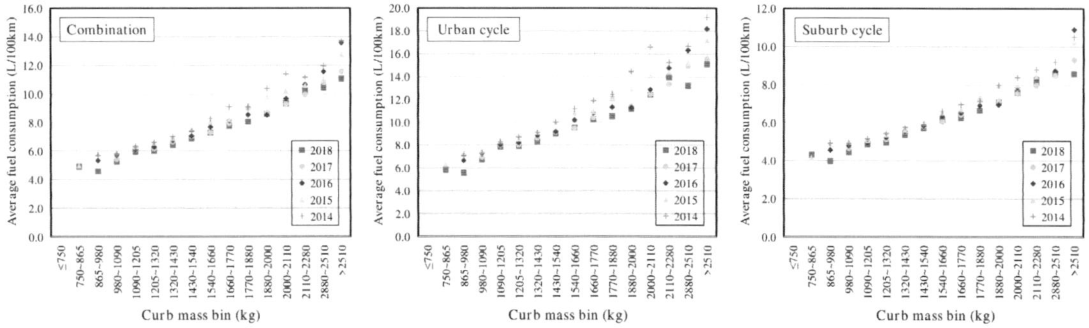

Figure 4. Variations of the average fuel consumption under each curb mass bin (Unban cycle, suburb cycle and NEDC)

3.2. Effects of turbocharge on fuel consumption reduction

As one of the main energy saving technologies, the main function of turbocharge is to compress the inlet charge before entering into the cylinder resulting in the increase of the output power and torque in order to realize the downsizing of the gasoline engine. Due to this advantage, turbocharge engines have been widely used in the gasoline passenger cars as shown in Figure 5. The model proportion of the gasoline passenger cars with turbocharge engines have more than doubled since 2014, and accounts for 67.8% in 2018.

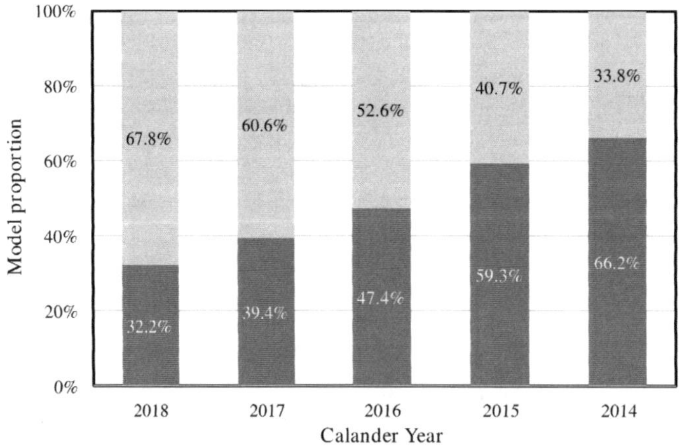

Figure 5. Model proportion of gasoline passenger car with turbocharge engines (green) and naturally aspirated engines (red)

In order to estimate the effect of curb mass on the fuel consumption, the fuel consumption per ton (L/(100km·t)) has been introduced which is the ratio between fuel consumption and the curb mass. Figure 6 showed the variations of the fuel consumption for the gasoline passenger cars with and without turbocharge engines. Results showed that the reduction of the average fuel consumption were 7.9% and 9.6% under NEDC for naturally aspirated engines and turbocharge engines in 2018 compared to 2014, and the annual decreasing rates were 2.0% per year and 2.4% per year respectively. It worth noting that the turbocharge engines' decreasing trend in fuel consumption was more obvious under urban cycle, and the annual decreasing rate reached 2.9% per year. This indicated development of passenger cars with turbocharge engines had a more significant effect on the fuel consumption reduction especially under urban cycle. Besides, the average fuel consumption of vehicles with turbocharge engines was 9.2% lower than naturally aspirated engines under NEDC cycle.

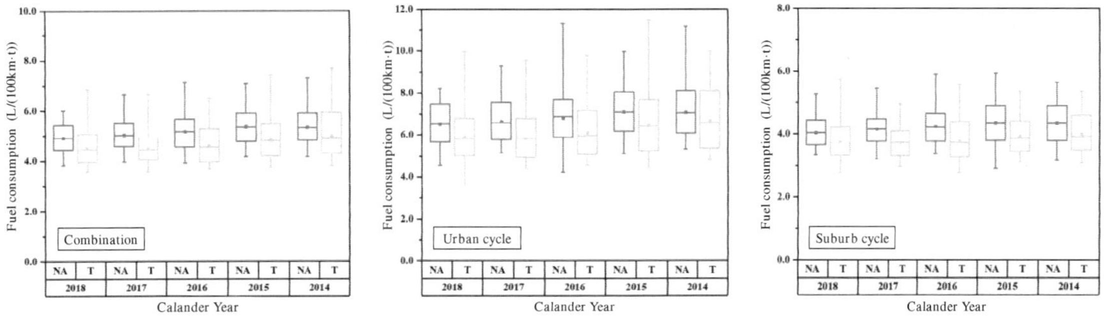

Figure 6. Variations of fuel consumption of gasoline passenger car with turbocharge engines and naturally aspirated engines under unban cycle, suburb cycle and NEDC

3.3. Effects of advanced transmission on fuel consumption reduction

In general, there are mainly 4 types of transmission used in China's gasoline passenger cars including MT, AT, CVT and DCT, and the model proportion of each transmission is shown in Figure 7. It could be seen that the proportions of the CVT and DCT increased sharply in recent years, especially DCT which have more than doubled since 2014. On the contrary, the proportions of MT and AT declined, but were still the mainstream. Since there was no fixed gear for CVT compared to other transmissions, this ensured the gasoline engines could always operate in the low fuel consumption region resulting in low fuel consumption. The energy saving advantage of DCT was mainly attributed to the application of 2 clutch discs sets instead of the hydraulic torque converter for shifting resulting in the continuous power and torque output of the gasoline engine.

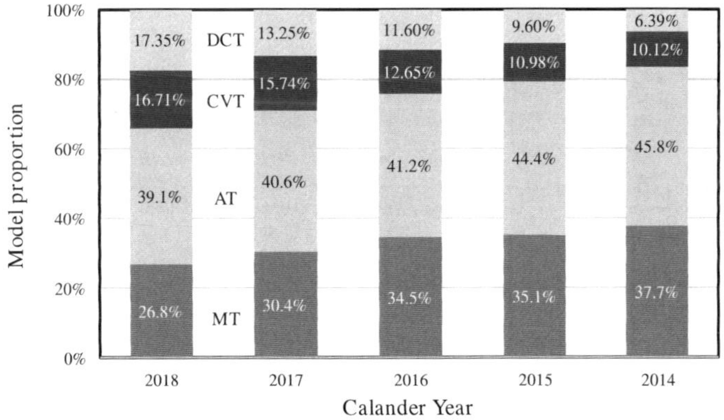

Figure 7. Model proportion of gasoline passenger car with MT, AT, CVT and DCT

Figure 8 showed the variations of the fuel consumption for the gasoline passenger cars with different transmission types. Results showed that the reductions of the average fuel consumption were 7.9%, 14.2%, 5.3% and 5.3% under NEDC for MT, AT, CVT and DCT in 2018 compared to 2014 respectively, the annual decreasing rate were 1.98% per year, 3.55% per year, 1.33% per year and 1.33% per year. It worth noting that the AT's decreasing trend in fuel consumption was more obvious under urban cycle, and the annual decreasing rate reached 4.1% per year. This indicated that the development of passenger cars with AT had the most significant effect on the fuel consumption reduction especially under urban cycle. The reason for the fuel consumption improvement was mainly contributed to the wide applications of AT with 7, 8 and even 9 gears. Besides, vehicles with DCT obtained the lowest fuel consumption under either cycle compared to other transmissions. For other transmissions, the average fuel consumption of CVT was 4.0% lower than MT and 3.2% lower than AT under urban cycle, which was due to the higher engine operating efficiency under the low speed and low load operating points. However, this advantage in fuel consumption weakened or even disappeared under suburb cycle, and the average fuel consumption of CVT was 2.6% lower than MT and 4.3% higher than AT. This was because that gasoline engines always operating in high efficiency region under the high speed and high load operating points no matter what the transmission type was, but the disadvantage of low transmission efficiency for CVT became prominent resulting in higher fuel consumption.

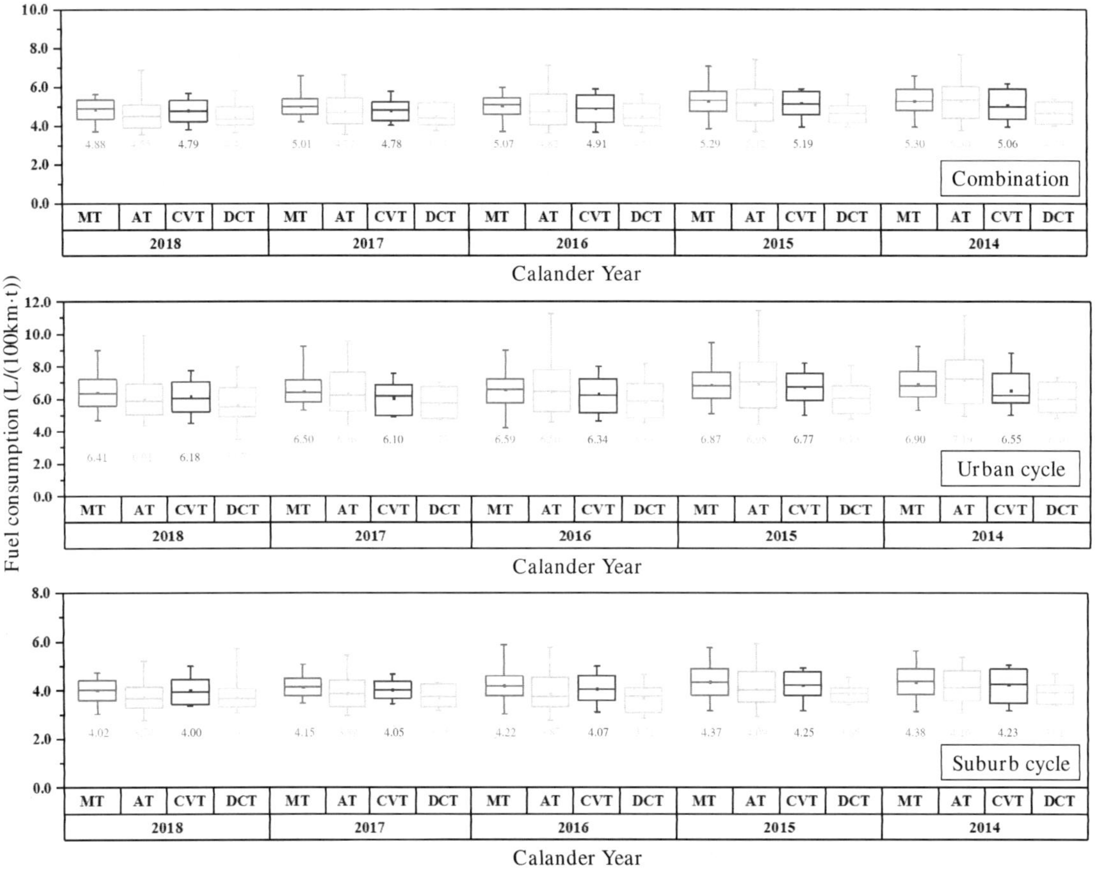

Figure 8. Variations of fuel consumption of gasoline passenger car with MT, AT, CVT and DCT under unban cycle, suburb cycle and NEDC

4. Conclusion

In this paper, the development of China's gasoline passenger car fuel consumption standards has been introduced. Moreover, the declared fuel consumption data of over 4800 gasoline passenger car models produced or imported in recent 5 years (2014-2018) has been analyzed, The development trends of the fuel consumption have been investigated first, and the effects of typical energy saving technologies including turbocharge and advanced transmission on the fuel consumption reduction have also been discussed. The conclusions can be reached as the following.

1. China's standards related to the fuel consumption of passenger cars went through five phases from 2004 to 2019 including the update of the per-vehicle fuel consumption limits and CAFC targets.
2. The average fuel consumption of China's passenger cars in 2018 was 7.13 L/100km, which was 6.4% lower than 2014, and the annual decreasing rate was about 1.6% per year. In addition, fuel consumption

decreasing under the urban cycle was the most obvious whose annual decreasing rate reached 1.98% per year.
3. The fuel consumption of the passenger cars increased linearly with the increase of curb mass, and the fuel consumption reduction of each year was more apparent for the gasoline passenger cars with higher curb mass.
4. The average fuel consumption of vehicles with turbocharge engines was 9.2% lower than naturally aspirated engines. The annual decreasing rate of fuel consumption of the turbocharge engines were also higher, about 2.4% per year, which indicated the development of passenger cars with turbocharge engines had a more significant effect on the fuel consumption reduction.
5. The average fuel consumption of vehicles with DCT was the lowest among 4 types transmission. The annual decreasing rate of the fuel consumption of AT was the highest, about 3.6% per year, which indicated that the development of passenger cars with AT had a more significant effect on the fuel consumption reduction.

5. References

[1] http://finance.sina.com.cn/china/gncj/2020-01-07/doc-iihnzahk2552277.shtml (in Chinese)
[2] https://baijiahao.baidu.com/s?id=1655661258154198782&wfr=spider&for=pc (in Chinese)
[3] The State Council of the People's Republic of China. 2012 Announcement No. 2012-22: Energy-saving and New Energy Vehicles Industry Strategic Planing (2012-2020) http://www.gov.cn/zwgk/2012-07/09/content_2179032.htm. (in Chinese)
[4] Standardization Administration of China. 2014 Fuel consumption limits for passenger car (GB 19578-2014) (in Chinese)
[5] Standardization Administration of China. 2019 Fuel Consumption Evaluation Methods and Targets for Passenger Cars (GB 27999-2019) (in Chinese)
[6] Patil C, Varade S, Wadkar S. 2017 A review of engine downsizing and its effects *International Journal of Current Engineering and Technology* **7(7)**: 319-324.
[7] de Salvo Junior O, de Almeida F G V. 2019 Influence of technologies on energy efficiency results of official Brazilian tests of vehicle energy consumption *Applied energy* **241**: 98-112.

Acknowledgments

This study is supported by the National Key R&D Program of China (2018YFB0106402 & 2018YFB0106404) and Research and Development of Chinese New Energy Automobile Products Test Driving Cycles (CATC).

IOP Conf. Series: Earth and Environmental Science **728** (2021) 012009 doi:10.1088/1755-1315/728/1/012009

Oxidative Desulfurization of Diesel from the Topping Plant using 1-Butyl-3-Methylimidazolium Hydrogen Sulfate [Bmim][HSO_4] as Catalyst

Sara Timbila[1], Caterine Donoso[1] and Francisco Mejía[2]

1 Departamento de Ciencias de la Energía y Mecánica, Universidad de las Fuerzas
2 Armadas ESPE, Av. General. Rumiñahui s/n, Sangolquí, P.O. Box 171-5-231B, Ecuador
2 Unidad de Operaciones, Petroamazonas E.P.
E-mail: iq.donosoc@gmail.com

Abstract. This investigation was carried out with the purpose of reducing the sulphur content in diesel from the topping plant and avoiding air pollution, due to the presence of SO_x in the emissions, at the same time reduced the corrosion rate in the equipment, machinery or mechanical components of the industry. An oxidative desulphurization (ODS) process was carried out with ionic liquid (1-butyl-3-methyl imidazolium hydrogen sulfate [BMIM][HSO_4]) that serves as a solvent and catalyst. H_2O_2 (30 % wt) was used as an oxidizing agent, because it is benign to the environment. A maximum of 43.14% removal was obtained at a volumetric ratio V_{IL}/V_{diesel} (1:1), molar ratio H_2O_2/ S (5) and reaction time (2 hours). In addition, the IL was evaluated after subjecting it to a regeneration process, being able to recycle it only once until it is saturated again; also the physical-chemical properties of diesel were studied before and after of exposed it to the ODS process.

1. Introduction

Desulphurization of fuel oils is an important process in oil refining, and hydrodesulphurization (HDS) is considered the traditional process widely used in industry. However, growing environmental concerns, legal limits imposed on sulphur content in fuels, and the ineffectiveness of HDS to remove the aromatic structure of sulphur compounds have led researchers to seek alternative methods of desulphurization. Methods such as oxidation, adsorption, extraction and bioprocesses have been investigated, and oxidative desulphurization (ODS) is considered the most competitive [1].

In the ODS, the sulphur compounds, such as dibenzothiophene (DBT) and its derivatives can be almost completely removed from diesel fuel by oxidation to their corresponding sulfones and subsequent extraction of these by some polar solvents such as: N-Methyl Pyrrolidone (NMP), N, N-dimethylformamide (DMF), Dimethyl Sulfoxide (DMSO), methanol (MeOH) and acetonitrile, in a simple process under less severe conditions, with low equipment and operating costs. Studies on ODS using IL have been reported recently in literatures. It has been observed that ODS using IL which served as both extractants and catalysts can effectively remove those HDS-immune cyclic S-compounds and reduce the S-component in fuel oils [2].

By means of the conventional oxidative desulphurization process, sulphur removal efficiency greater than 90% can be obtained by using a solution of hydrogen peroxide (H_2O_2) as an oxidant, which is combined with catalysts of the complex type of acetic acid, peroxotungsten or peroxomolibdene.

However, the use of these catalysts raises some problems such as the difficulty in separating or regenerating them, as well as in the contamination of the fuel, so alternative catalysts are investigated such as ionic liquids (IL) for use in this process [2].
Most works, were focused on the investigation of model diesel fuel. However, ODS of the real oil is rarely studied and the desulphurization efficiency is not ideal. Therefore, for this study, real diesel produced in the topping plant was used, this product content 3000 ppm S approximately. In previous work, with real diesel of 97 ppm S, 85.5 % S-removal efficiency after one stage of ODS with IL 1-butyl-3-methyl imidazolium hydrogen sulfate [BMIM][HSO_4], and other studies with IL Brønsted-Lewis acid zinc chloride N-methylpyrrolidonium ([Hnmp] Cl / ($ZnCl_2$), for FCC real diesel fuel, removal efficiency is less than 38% in a one stage and can reach 83 % after five stages [3].
In ODS study conducted at temperatures of 25 °C, 40 °C and 60 °C, room temperature was considered the most optimal, because a high temperature decreases the viscosity of the IL, which is favorable by allowing a complete mixture between this and the fuel, however, there is a rapid decomposition of H_2O_2. When the reaction temperature increased from 25 °C to 60 °C, the removal of sulphur from DBT decreased considerably from 99.6% to 64.6 %; therefore, the optimal reaction temperature was adjusted to room temperature [4].

2. Experimental Section

2.1. Reagents
1-Butyl-3-methyl imidazolium hydrogen sulfate, ≥ 95% purity imported and supplied by SIGMA-ALDRICH, Hydrogen Peroxide (30%), 99.7% purity Diethyl Ether.

2.2. Equipment
To measure the sulphur content, an X-ray fluorescence spectrophotometer, HORIBA brand, model: SLFA-20 / SLFA-2100 / SLFA-2800 was used, and to measure the other properties of diesel, the equipment of the Refinery Laboratory was used of Block 56 / Petroamazonas EP.

2.3. Oxidative Desulphurization
The ODS reaction was carried out in a 50 ml round bottom flask, IL and 30 wt % H_2O_2 solution were put into bottom, the diesel fuel was added in the required proportion and the mix were magnetically stirred at some temperature(25 °C) for a certain time (1, 2 or 3 hours). After a certain time of oxidation, the mixture was laid aside for 3 hours for phase separation. The diesel phase was analyzed by X-ray fluorescence spectrophotometer.

2.4. Regeneration of IL
The regeneration of IL is of vital importance for the industrial application, in this case, the process to regenerate the IL was carried out as follows, the IL phase was treated with rotary evaporation to remove impurities and regenerate the IL to reuse it in ODS.

Distilled water was first added in the same volume proportion, as the extract phase obtained. The mixture was produced stirred for 10 minutes at 1000 rpm, at temperature of 25 ºC. Thereafter, the mixture was treated at 105 ºC using a rotary evaporator at a vacuum of 420 mbar for three hours to remove the water. The regenerated IL, was washed three times with diethyl ether with a volume equal to that of the extract obtained, to later perform the separation in a rotary evaporator and recover the IL for recycling [5]. A summary scheme of what was done in the laboratory can be seen in figure 1.

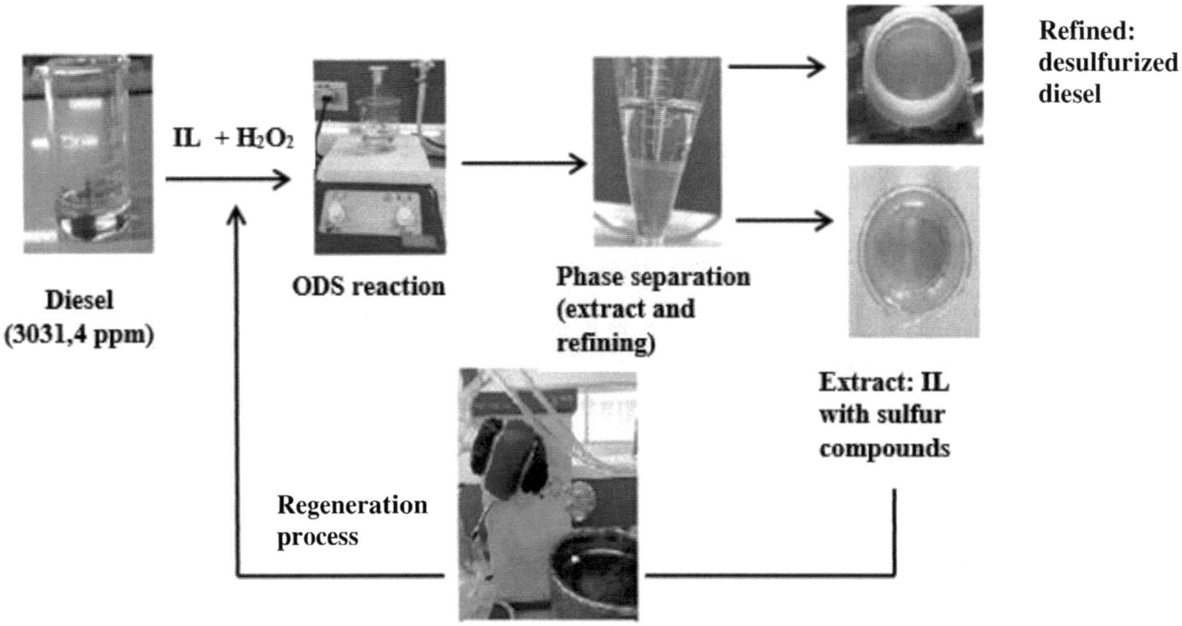

Figure 1. Schematic representation of the process of desulphurization of diesel with IL.

3. Results and Discussion

3.1. Effect of the volumetric ratio VIL/Vdiesel

The effect of volumetric ratio IL / fuel of 1/3, 1/2 and 1/1 were evaluated, as can be seen in figure 2, the volumetric ratio that has the highest proportion is the one with the best performance, either that work with a molar ratio of 3, 4 or 5. The highest S-removal efficiency of 43.14 % is obtained when the volumetric ratio of IL/ fuel was 1/1.

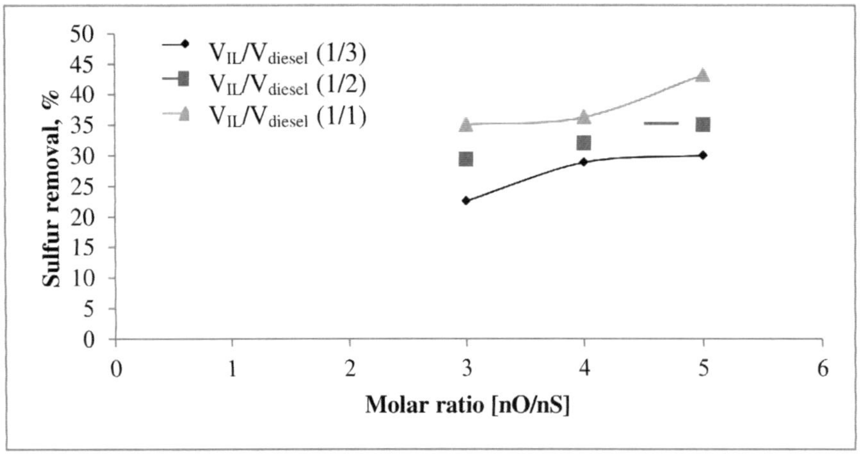

Figure 2. Effect of the volumetric ratio V_{IL}/V_{diesel}.

3.2. Effect of molar ratio nO/nS

According to the stoichiometric reaction, to oxidize the sulphur-containing compounds to their corresponding sulfones using hydrogen peroxide, 2 mol of hydrogen peroxide is consumed per 1 mol of sulphur-containing compound. As show in figure 3, the sulphur removal was 7.51% for the molar ratio of 2, this can be attributed to the fact that there is not enough H_2O_2 to completely oxidize the sulphur compounds. The removal efficiency increased to 28.85 % when nO/nS was 5. This effect is assigned to the increase in the concentration of H_2O_2 resulting in a greater possibility of oxidizing sulphur compounds. However, when working with nO / nS equal to 8 and 10, the efficiency decreased, it was considered that at molar ratios greater than 5, there is no longer a significant effect of the oxidant, an excess of H_2O_2 introduces water into the reaction environment and dilutes the IL, so it negatively influences the ODS process [6].

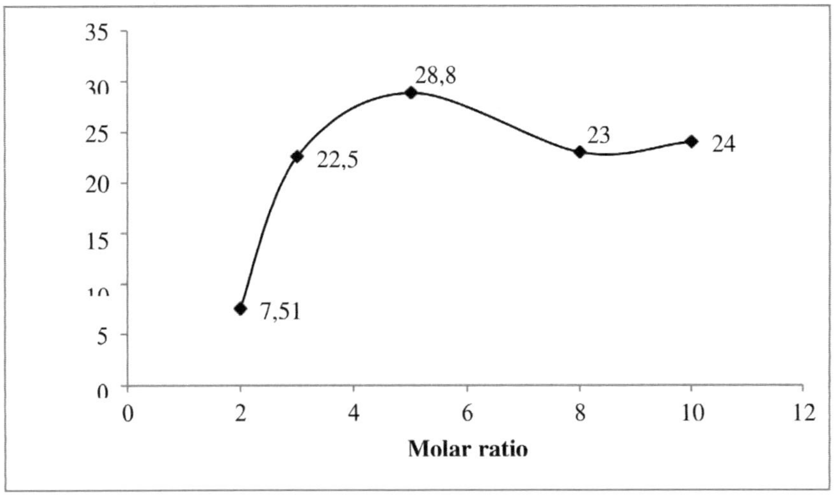

Figure 3. Effect of the oxidant on the removal efficiency of S (Cond: T (25° C) , stirring for 2 hours, V_{IL} / V_{diesel} = 1: 3).

3.3. Effect of reaction time

The effect of the reaction time (30 min, 1 h, 2 h, 3 h and 5 h) were evaluated at the conditions: volumetric ratio V_{IL} / V_{diesel} (1: 3), molar ratio nO / nS (5), temperature (25 °C) and 900 rpm stirring. According to figure 4 here are approximately equal removal values between 1, 2 and 3 hours. However, at 30 min and greater than 5 hours the removal efficiency decreases, in the latter case, the increase in the contact time of the phases generated an increase in the solubility of the IL in the fuel, which was verified with the sulphur content result equal to 3535.9 ppm.

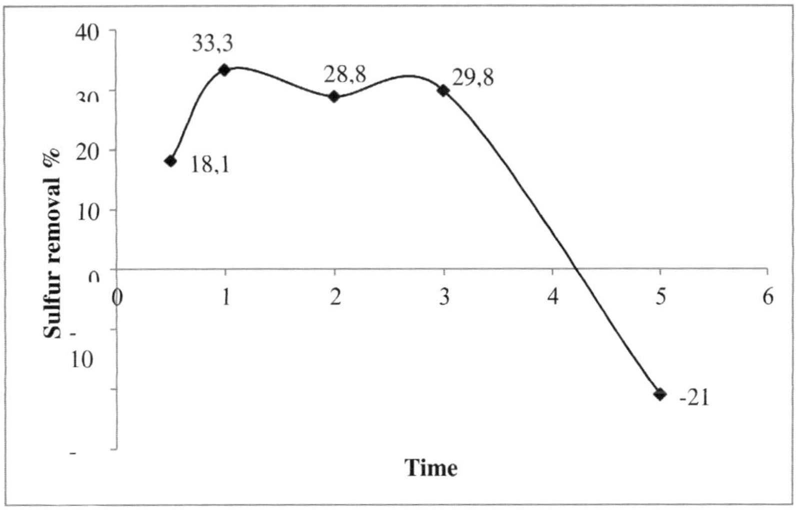

Figure 4. Effect of reaction time on the removal efficiency.

3.4. Effect of regenerated IL

To study the extracting effect of the regenerated IL [BMIM] [HSO_4], was worked to the best conditions of the factorial design, at volumetric ratio V_{IL} / V_{diesel} equal to 1: 1, molar ratio nO / nS equal to 5, time of stirring for 2 hours, temperature of 25 ° C in a single stage and with the same load of feeding. Three regeneration cycles were performed, considering that each cycle corresponds to a stage of extraction and regeneration. The results of the efficiencies for the three cycles are shown in table 1.

Table 1. Regeneration of IL saturated.

Cycle	Extraction efficiency
0	43.13
1	34.51
2	19.67
3	5.70

According to the results in table 1, there is no positive effect when performing this process, since in the first regeneration cycle a drop of almost 9% in sulphur removal efficiency is observed. For the second and third regeneration cycle there are minor efficiencies, that is, the IL is very saturated. These results indicate that the IL [BMIM] [HSO_4] can be regenerated only once so that there is no appreciable loss in the percentage of removal obtained.

3.5. Partition coefficients (Kd)

These values are provided to specify the capacity for extracting the IL. The partition coefficient is a key parameter to determine the solvent extraction capacity in an industrial extraction process. The higher the partition coefficient, the better the desulphurization performance of an IL [7]. In figure 5 this trend is observed, at a lower K_d there is a lower percentage of sulphur removal. Regardless of the time and amount of oxidant that has been used in the test, this tendency is observed for each of the volumetric ratios tested.

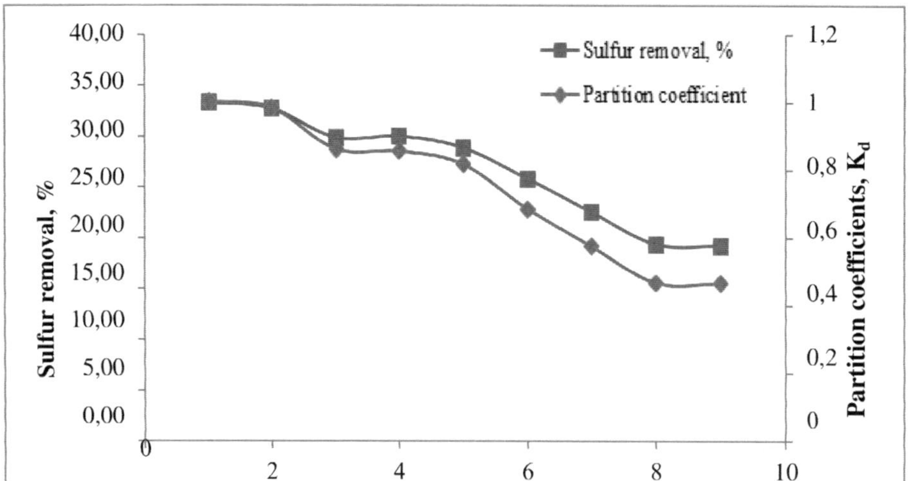

Figure 5. S removal efficiency and partition coefficient, K_d (Cond: V_{IL} / V_{diesel} = 1: 3).

3.6. Diesel properties before and after process ODS

After the diesel is subjected to an ODS process with IL, it is expected that its quality will not be affected, as showed in table 2; however, there are certain variations, for example, the flash point increased to 72.42 °C. The viscosity had a minimal change; however, the result is favorable in terms of having a better-quality fuel. The sulphur content that this fuel has, decreased by almost 1500 ppm when using the ODS process described in the present investigation.

Properties such as copper sheet corrosion are not modified because the diesel fuel itself is not corrosive to copper or its alloys. And the Cetane Index Calculated did not suffer a negative change after the ODS process with IL. In general, most of the properties analyzed for this study were not seen affected after the fuel was in contact with the IL in the ODS process, indicating that the product refined with IL does not suffer alterations in its physical-chemical properties [8].

Table 2. Properties of diesel before and after of exposed it to the ODS process.

Parameter	Diesel before ODS	Diesel after ODS
°API at 15°C	35.04	35.54
Density at 15 ° C (g mL^{-1})	0.8494	0.8471
Flash point (°C)	66.33	72.42
Distillation temperature (90%) (°C)	344.3	343.5
Kinematic Viscosity at 37.8 ° C (cSt)	3.311	3.28
Sulphur (%wt)	0.3031	0.1346
Corrosion to copper foil	No. 1	No. 1
Cetane Index Calculated	49.6	51.3

4. Conclusion

The ODS process of diesel fuel with an initial sulphur content of 0.3031% wt with IL was worked at room temperature (25 ° C), at the volumetric ratio of V_{IL} / V_{diesel} to 1: 1, at H_2O_2 oxidant (nO / nS to 5) and at a reaction time of 2 hours to reach 43.14% sulphur removal. All the physicochemical properties of diesel evaluated in the present investigation, among them the calculated cetane index and viscosity, were not significantly modified after the ODS process with IL, so the quality of the same improves due to the decrease in sulphur compounds. The regeneration of the IL [BMIM] [HSO_4] saturated with sulphur compounds were not possible, since during the first reuse cycle the removal percentage decreased by 9 %, that is, the IL did not fully recover its catalytic activity and solvent, these is the principle reason by the ODS with IL must be a complementary and subsequent process to conventional HDS. The cost of reagents used into desulphurize diesel fuel of 0.3% to 0.1% wt, without pretreatment of HDS was 100 USD per gram of sulphur removed, at the laboratory conditions described in this study, however is necessary a total economic analysis considered energy consumption, greenhouse gas emissions and operational costs.

5. References

[1] Gao, S., Li, J., Chen, X., Abdeltawab, A. A., Yakout, S. M., & Yu, G. (2018). *A combination desulfurization method for diesel fuel: Oxidation by ionic liquid with extraction by solvent*. Fuel, 224, 545-551.

[2] Chen, X., Song, D., Asumana, C., & Yu, G. (2012). *Deep oxidative desulfurization of diesel fuels by Lewis acidic ionic liquids based on 1-n-butyl-3-methylimidazolium metal chloride*. Journal of Molecular Catalysis A: Chemical, 359, 8-13.

[3] Chen, X., Guan, Y., Abdeltawab, A. A., Al-Deyab, S. S., Yuan, X., Wang, C., & Yu, G. (2015). *Using functional acidic ionic liquids as both extractant and catalyst in oxidative desulfurization of diesel fuel: An investigation of real feedstock*. Fuel, 146, 6-12.

[4] Gao, H., Guo, C., Xing, J., Zhao, J., & Liu, H. (2010). *Extraction and oxidative desulfurization of diesel fuel catalyzed by a Brønsted acidic ionic liquid at room temperature*. Green chemistry, 12(7), 1220-1224.

[5] Valenzuela, C., Donoso, C., Guzmán, L. (2020). *Extraction of sulfur from commercial gasoline using 1-butyl-3-methylimidazolium tetrafluoroborate [BMIM][BF_4] as the extraction solvent.* Key Engineering Materials, 834, 42-48.

[6] Safa, M., Mokhtarani, B., Mortaheb, H. R., & Tabar Heidar, K. (2016). *Oxidative desulfurization of model diesel using ionic liquid 1-octyl-3-methylimidazolium hydrogen sulfate: an investigation of the ultrasonic irradiation effect on performance*. Energy & Fuels, 30(12), 10909-10916.

[7] Dharaskar, S. A., Deshmukh, S. K., Bhuyar, K. D., & Wasewar, K. L. (2015). *Ionic Liquids: As Energy Efficient Solvent For The Extractive Desulfurization of Liquid Fuels*. In 3rd International Conference on Chemical, Agricultural and Medical Sciences (CAMS-2015) (pp. 10-11).

[8] Timbila, S. E., (2019). *Desulfuración oxidativa del diésel procedente de la planta topping de la Refinería de Lago Agrio empleando sulfato de hidrógeno 1-butil-3_ metilimidazolio [BMIM][HSO$_4$] como catalizador* (Ecuador: Universidad de las Fuerzas Armadas ESPE) p 111-112.

Research on Advanced Available Techniques of Air pollution Prevention and Control in Building Ceramics Industry in China

H Su[1], H M Li[1], X C Wang[2], Z S Wang[3], Y F Tan[3*]

1 Capital University of Economics and Business, School of Safety and Environmental Engineering, Beijing100071, China
2 Institute of Technical Information for Building Materials Industry, Beijing100024, China
3 Chinese Research Academy of Environmental Sciences, Beijing100012, China
E-mail: tanyufei_es@163.com

Abstract. The flue gas emission of the building ceramics industry accounts for 85% of the whole ceramic industry in China, and it is the critical sector of air pollution control in the ceramic industry. The advanced available techniques (AAT) of air pollution prevention and control are of considerable significance to promote the upgrade of pollution prevention measures and technological progress. The paper studied and established the method of screening and evaluating the AAT for the prevention and control of air pollution in the building ceramics industry. Through collection and analysis of three normal pollutant emission concentrations of ten available techniques (AT), to select the AAT that meet at least one pollutant of all the plants was lower than 50% of the national standard emission concentration threshold in China. Based on the experience of the European Union and the comprehensive evaluation of industry associations, research institutions, and pollution control experts, five AAT were selected. The findings will provide an essential reference for the building ceramics industry to select advanced techniques, and provide significant technical support for the revision of the ceramic industrial pollutant emission standard for and selection of pollution prevention and control techniques (PPT and PCT) for pollutant discharge permission management.

1. Introduction

In recent years, China's building ceramics industry has developed rapidly. In 2016, the output reached 110.76 billion m^2, accounting for two-thirds of the world's total output [1]. While it has made significant contributions to the economy, it has also brought pressure on environmental protection and energy consumption. China's ceramic tile fume emissions account for about 85% of China's ceramic industry, and it is a crucial sector in the ceramic industry for air pollution control. Ceramic tile flue gas is generated in kiln and spray drying tower, and its primary pollutants are particulate matter, sulfur dioxide, and nitrogen oxides. Advanced available techniques (AAT) for the prevention and control of air pollution in ceramic tile have an essential role in mitigating the generation and discharge of pollutants from ceramic tile industrial plants and thereby reducing their impacts on the environment. In 2006, the European Union issued the "Reference Document on BAT in the Ceramic Manufacturing Industry" [2], which made high-level environmental protection of the ceramic industry in the most effective way. In 2018, the Ministry of Ecology and Environment of China issued the "Guideline on available techniques of pollution prevention and control for ceramics manufacturing industry" (HJ 2304-2018) [3], and recommended 10 available technique (AT) combinations, but did not recommend AAT which can

IOP Conf. Series: Earth and Environmental Science **728** (2021) 012010 doi:10.1088/1755-1315/728/1/012010

achieve higher level of environmental economic benefits. With China's "13th Five-Year Plan for Ecological Environment Protection" [4], pollution prevention battle, pollutant permit system and other stricter requirements for air pollution emissions, screening a series of AAT for the protection and prevention of atmospheric pollution in the ceramics tile industry with higher environmental-economic benefits are of great significance to promote the ceramic tile plants upgrading pollution prevention and control measures and technological progress.

2. Overview of available techniques for air pollution control in the building ceramics industry

During the compilation process of HJ 2304-2018, through literature and research, expert discussions, field investigations, on-site tests, etc., ten technique combinations including prevention techniques, kiln gas treatment techniques, and spray drying tower gas treatment techniques were selected and specified [5]. Table 1 shows the AT for pollution Prevention and Control in building ceramics industry recommended in HJ 2304-2018. Among them, 1 to 7 are the AT for the treatment of kiln flue gas and spray drying tower gas respectively, and 8 to 10 are the combining treatment techniques for kiln flue gas and spray drying tower flue gas. These ten recommended techniques have continuous and stable operation cases in China.

3. Methods for screening and evaluation advanced available techniques

3.1. Research on screening methods

The AAT in HJ 2300-2018 is defined as “Among the available techniques for pollution prevention and control, the technique that makes the discharge of at least one major pollutant stably below 70% of the national pollutant emission standard limit" [6]. According to the survey, most of the ceramic tile plants in China can achieve this requirement by adopting AT. Therefore, the requirement of at least one pollutant below the standard limit of 70% is loose for Chinese building ceramics industry. According to this requirement, it is also impossible to screen out AAT for the building ceramics industry.

IPPC 96/61/CE [7] and Directive 2010/75/EC [8] issued by The European Union all mentioned Best Available Technique (BAT). "Best" means that the technique considered is the "most effective in achieving a generally high level of protection of the environment as a whole" [7], and introduced the BAT evaluation criteria, including the reduction of waste generation, the use of low-harm raw materials, scale applications, representing advanced technique and development direction. The selection and evaluation of BAT have strict procedures. The determination method is mainly through technical information exchange of government departments, non-government departments, industry associations, research institutions, etc., rather than quantitative screening methods.

Table 1. AT for pollution Prevention and Control in building ceramics industry [3].

AT	Prevention Techniques	Kiln Gas Control Techniques	Spray drying tower Gas Control Techniques
1	①Raw material PCT (low sulfur materials) +②Kiln gas PPT +③Spray drying tower coal-water slurry hot blast stove coal-based fuel high-temperature sulfur fixation technique	Wet desulfurization (lime-gypsum or sodium-alkali) collaborative dust removal	①SNCR in hot air stove +② bag-type dust removal +③Wet desulfurization (lime-gypsum or sodium-alkali) collaborative dust removal
2	①Raw material PCT (low sulfur materials) +②Kiln gas PPT +③Spray drying tower coal water slurry hot blast stove (coal-based fuel high-temperature sulfur fixation + low nitrogen combustion) or generator gas hot blast stove low nitrogen combustion technique	Wet desulfurization (lime-gypsum or sodium-alkali) collaborative dust removal	①Bag-type dust removal +② Wet desulfurization (lime-gypsum or sodium-alkali) collaborative dust removal
3	①Raw material PCT (low sulfur materials) +②Kiln gas PPT+③Low-nitrogen combustion technique of methane hot blast stove in spray drying tower	Wet desulfurization (lime-gypsum or sodium-alkali) collaborative dust removal	①Bag-type dust removal + ② Spray-type dust removal

AT	Prevention Techniques	Kiln Gas Control Techniques	Spray drying tower Gas Control Techniques
4	①Raw material PCT (low sulfur materials) + ② Kiln gas PPT +③Spray drying tower pulverized coal chain type hot blast stove low sulfur fuel technique(low sulfur coal) +④Spray drying tower pulverized coal chain type hot blast stove coal- based fuel high-temperature sulfur fixation +⑤ Spray drying tower pulverized coal chain type hot blast stove low nitrogen combustion	Wet desulfurization (lime-gypsum or sodium-alkali) collaborative dust removal	①Bag-type dust removal + ② Spray-type dust removal
5		Wet desulfurization (lime-gypsum or sodium-alkali) collaborative dust removal	①Cyclone-type dust removal +②Spray-type dust removal + ③ Wet electric-type dust removal
6	①Raw material PCT (low sulfur materials) +②Kiln gas PPT + ③ High-temperature sulfur-fixing technique for coal-based fuel in the coal-fired hot-blast stove of spray drying tower	Kiln gas wet method for multi-pollutant collaborative control technique	①SNCR in hot air stove +② bag-type dust removal +③Wet desulfurization (lime-gypsum or sodium-alkali) collaborative dust removal
7	①Raw material PCT (low sulfur materials) +②Kiln gas PPT+ ③ High-temperature sulfur-fixing technique for coal-based fuel in the coal-fired hot-blast stove of spray drying tower + Low-nitrogen combustion technique for the coal-fired hot-blast stove of spray drying tower	Kiln gas wet method for multi-pollutant collaborative control technique	①Bag-type dust removal + ② Wet desulfurization (lime-gypsum or sodium-alkali) collaborative dust removal
8		①SNCR denitration in hot air stove of spray drying tower + ② Bag-type dust removal of spray drying tower + ③ Cooperative dust removal of kiln gas and spray drying tower gas wet desulfurization (lime-gypsum or sodium-alkali)	
9	①Raw material PCT (low sulfur materials) +②Kiln gas PPT +③Low-sulfur fuel technique for hot blast stove of spray drying tower (low-sulfur coal) +④ High-temperature sulfur fixation technique for hot blast stove of spray drying tower for coal-based fuel	①SNCR denitration in hot air stove of spray drying tower + ② Bag-type dust removal of spray drying tower + ③ Cooperative dust removal of kiln gas and spray drying tower gas wet desulfurization (lime-gypsum or sodium-alkali)+④ Wet electric-type dust removal	
10		①SNCR denitration in hot air stove of spray drying tower + ②Cyclone dust removal of spray drying tower gas +③Semi-dry desulfurization and desulfurization of kiln gas and spray drying tower gas circulating fluidized bed	

The Clean Air Act of 1990 issued by the United States stipulated that New Source Performance Standards (NSPS) should be established based on Best Demonstrated Technique (BDT) [8], and National Emissions Standard for Hazardous Air Pollutants (NESHAP) should be based on maximum available control technique (MACT) [9]. Among them, the screening of BDT is determined through a comprehensive assessment of reduction levels, implementation costs, and demonstration verification, and the screening of MACT is determined by counting the emission levels that can be achieved by various control techniques. The MACT of the new source is the technique with the lowest emission level that can be achieved by the same source. The MACT of the existing source is the technique that can reach the control level of the best 12% pollution source (more than 30 pollution sources) or the average pollution control techniques that the top 5 of similar pollution sources can reach for pollution sources less than 30.

At present, there are few studies on screening methods of AAT for pollution prevention in China, and there are no scientific, reasonable, objective, and fair evaluation index system and evaluation methods. According to HJ 2300-2018, determining the definition of AAT refers to the "Environmental Protection Tax Law of the People's Republic of China", it said " If the concentration value is lower than 30% of the national and local pollutant emission standards, the environmental protection tax shall be reduced by 75%", indicating that emissions stably below 70% of emission standards. The "Environmental Protection Tax Law" aims to promote plants to reduce the concentration of pollutant emissions effectively. To truly reflect the most economical and practical level that AAT can achieve, the screening research is still mainly based on the emission concentration levels that can be achieved by each technique. As the requirement of at least one pollutant lower than the standard limit of 70% is obviously loose for the building ceramics industry, this study refers to the "Environmental Protection Tax Law" which is " If the concentration value is lower than 50% of the national and local pollutant emission standards, the environmental protection tax shall be reduced by 50%". 50% below the emission limit can theoretically effectively to reduce the emission concentration and has a prominent green leading role. Therefore, this study takes "the emission concentration of at least one major pollutant below 50% of the emission standard" as the AAT screening standards. At the same time, they are combined with the experience of the European Union, the comprehensive evaluation of industry associations, research institutions, and pollution control experts to select AAT better.

3.2. Evaluation Methodology

At present, China has not established a sound technical and economic evaluation system for AAT. Descriptive indicators for BAT assessment in the European Union and the United States mainly include indicators such as resource consumption, energy consumption, pollution emissions, economic costs, environmental effects, social impacts, and case analysis, etc.. The United States technical assessment is mainly based on whether the pollutant emissions meet the emission limit standards, and the EU's indicator settings aim to be able to describe BAT parameters in detail. At present, China has carried out AT assessment studies for some key industries, including the Chinese thermal power industry [10], lead smelting industry [11], Vitamin C industry [12] and textile dyeing and finishing industry [13] . The technical evaluation of the study focuses on the four indicators of pollutant emissions, investment costs, economic costs, and technical-economic parameters. The operating costs include resource consumption and energy consumption. At the same time, the method of an expert-assisted comprehensive evaluation is adopted to compare and evaluate AAT comprehensively, and finally, screen out AAT that meet the status of industry development and the actual situation of the ceramic tile industry.

4. Selection of advanced available techniques

4.1. Analysis of Atmospheric Emission Data of Chinese Ceramic Tile plants

By collecting and analyzing the compliance monitoring data and technical cases of typical ceramic tile plants corresponding to ten available ceramic air PPT and PCT recommended by HJ 2304-2018. The concentration levels of particulate matter, sulfur dioxide and nitrogen oxides in the exhaust gas using various techniques are shown in Figure 1.

4.2. Selection of advanced available techniques

According to the screening method in 3.2, the number of plants and the proportions of particulate matter, sulfur dioxide, and nitrogen oxides for each AT, which are lower than 70% and 50% of the standard limits, can be counted. Table 2 shows the comparative analysis results of the emission concentrations of 10 AT and the standard emission concentration limits [14, 15]. All these ten techniques can meet the advanced available technical requirements of HJ 2300-2018, which "at least one major pollutant stably below 70% of the national pollutant emission standard limit". AT 3 and 9 can meet the emission concentration of the three pollutants of all plants below 50% of the standard emission limit. However, as for available technique 3, there are only two companies have collected emission monitoring data, and combined with a comprehensive evaluation by ceramics experts, whether this technique can stably fall below 50% of the standard emission limit requires further research and verification. Therefore, the available technique 9 screened out in this study has technologically advanced feasibility at the level of emission concentration and is the most advanced available technique. The AT 4, 5, 6, 7, 8, and 10 can meet the emission concentration of one pollutant of all plants below 50% of the standard emission limit. Combined with a comprehensive assessment of experts in the ceramic industry, the stability of AT 4, 6, and 7 also needs further study. Therefore, AT 5, 8, and 10 are selected as AAT. Besides, due to limited application plants, the SCR de-nitration technique is not included in the list of AT for pollution prevention and control of the ceramic manufacturing industry recommended by HJ 2300-2018. Still, this technique is mature in theory, and there are handful successful and stably cases in China. There are at least 3 plants in China adopted "raw material control technique (low sulfur materials) + kiln gas pollution prevention technique + dry powder milling technique + SCR denitration technique". The desulfurization rate is greater than 98%, the de-nitration rate is greater than 90%, and the dust removal rate is greater than 99%. Kiln gas emission monitoring report shows that particulate matter emission concentration is lower than 10mg / m^3, SO_2 emission concentration is lower than 20mg / m^3, NO_x emission concentration is lower than 50mg / m^3, and the emission concentration of all three pollutants is lower than 50% of the standard emission limit. Also, the SCR technique worked stably, and it is worth promoting in the building ceramics industry. Therefore, this technique should be added as a "complementary available technique" in the list of AAT.

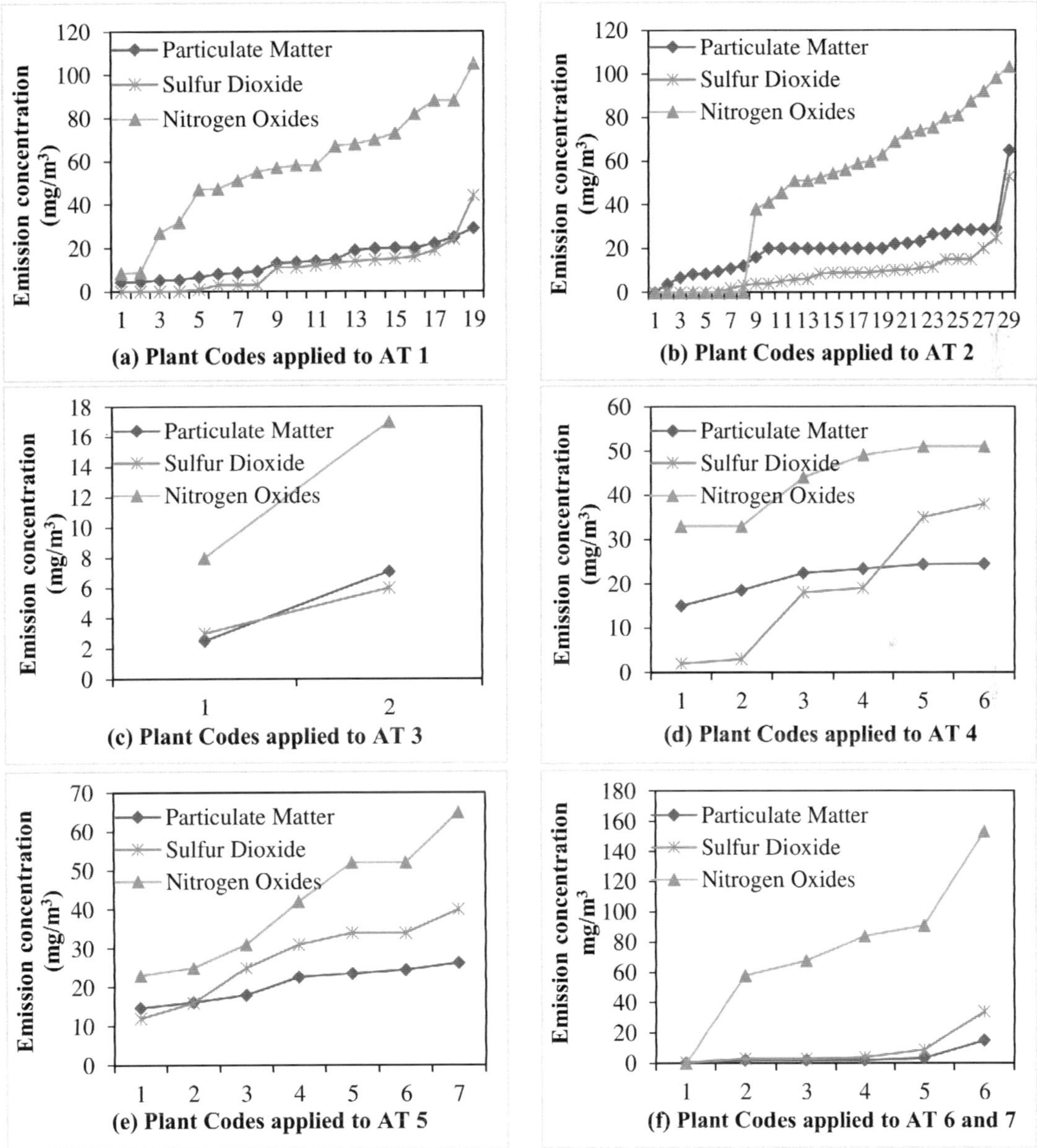

(a) Plant Codes applied to AT 1

(b) Plant Codes applied to AT 2

(c) Plant Codes applied to AT 3

(d) Plant Codes applied to AT 4

(e) Plant Codes applied to AT 5

(f) Plant Codes applied to AT 6 and 7

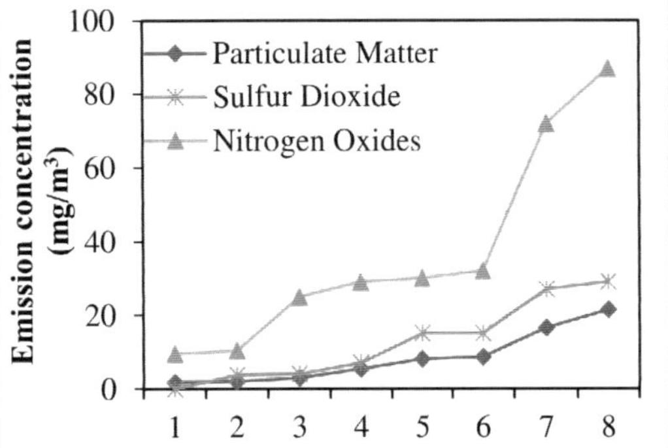

(g) Plant Codes applied to AT 8

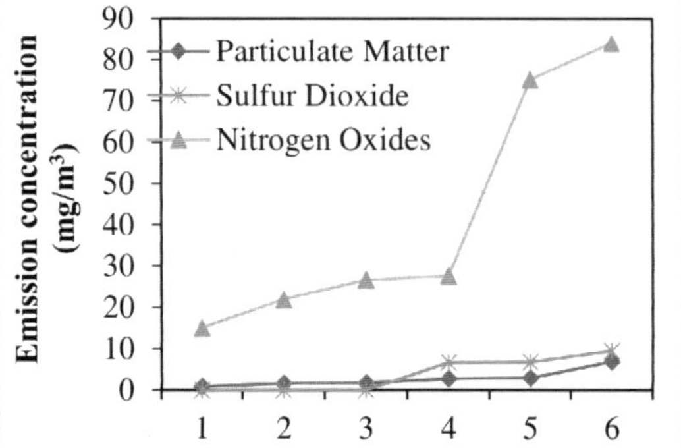

(h) Plant Codes applied to AT 9

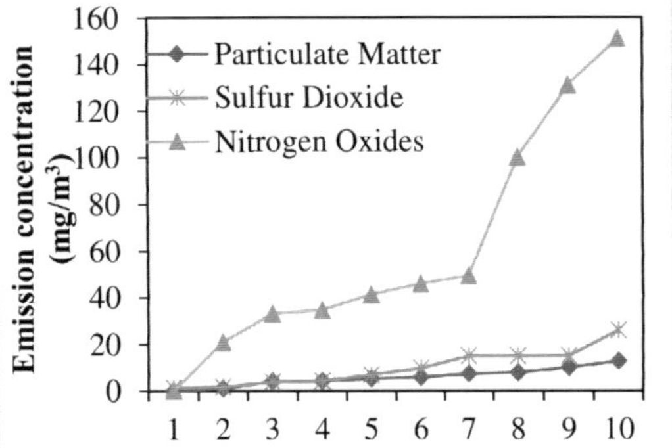

(i) Plant Codes applied to AT 10

Note: The X-axis of the graph (a) to (i) is different from each other because the number of plants applied to available technology (AT) 1 to 10 is different.

Figure 1. Plant Emission concentration level of ten available Techniques.

Table 2. Comparison of 10 AT emission concentration and the standard limit (SL).

AT	Type of Pollution	SL (mg/m^3)	Number of companies below 70% SL	Proportion (%)	Number of companies below 50% SL	Proportion (%)
	PM	30	15	78.95	12	63.16
1	SO_2	50	18	94.74	18	94.74
	NO_x	180	19	100	18	94.74
	PM	30	19	65.52	8	27.59
2	SO_2	50	28	96.55	28	96.55
	NO_x	180	29	100	26	89.66
	PM	30	2	100	2	100
3	SO_2	50	2	100	2	100
	NO_x	180	2	100	2	100
	PM	30	2	33.33	0	0.00
4	SO_2	50	5	83.33	4	66.67
	NO_x	180	6	100.00	6	100.00
	PM	30	3	42.86	1	14.29
5	SO_2	50	6	85.71	3	42.86
	NO_x	180	7	100.00	7	100.00
	PM	30	6	100.00	6	100.00
6 and 7	SO_2	50	6	100.00	5	83.33
	NO_x	180	5	83.33	4	66.67
8	PM	30	7	87.50	6	75.00
	SO_2	50	8	100.00	6	75.00

IOP Conf. Series: Earth and Environmental Science **728** (2021) 012010 doi:10.1088/1755-1315/728/1/012010

	NO_x	180	8	100.00	8	100.00
	PM	30	6	100.00	6	100.00
9	SO_2	50	6	100.00	6	100.00
	NO_x	180	6	100.00	6	100.00
	PM	30	10	100.00	10	100.00
10	SO_2	50	10	100.00	9	90.00
	NO_x	180	8	80.00	7	70.00

5. Technical and economic feasibility assessment

Through the collection of data on investment costs, operating costs, technical-economic parameters of AAT selected previously, and combined with relevant expert consultation and comprehensive evaluation, the technical-economic feasibility of AAT are evaluated. The evaluation results are shown in the table 3. As can be seen from Table 3, the investment cost and operating cost of the available technique 9 with the lowest emission concentration level is significantly higher than other techniques, while the second-AAT 5 has the lowest investment cost and more economical operating cost although the emission level is higher than other techniques. Also, "complementary AT" have the most economical operating costs. Plants can make reasonable choices based on local environmental management needs, pollutant discharge targets, and affordable economic costs.

Table 3. Statistical table of economic evaluation of AAT.

AT	Investment costs (million RMB)	Operating costs (RMB/day)	Technical-economic parameters (m^3/h)
9	13	10600	580000
8	8	6000	580000
10	12	6000	580000
5	2.81	3253.7	129882
Complementary	4.20	1400	250000

6. Conclusions

a) Based on the corresponding emission concentration level requirements of the "Environmental Protection Tax Law" for corporate tax reduction and exemption, from the 10 AT recommended by HJ2304-2018, select at least one pollutant that meets all plants' stable emission concentrations below 50 % of the standard limits, and draw on the EU experience combined with the comprehensive evaluation of industry associations, research institutions, and pollution control experts to screen out five AAT.

b) Based on the indicators of pollutant discharge, investment cost, economic cost, and technical-economic parameters, the technical-economic feasibility of 5 AAT are evaluated. The results show that the investment and operating costs of available technique 9 with the most substantial reduction in emissions are significantly higher than other techniques, while available technique 5 has the lowest investment costs and lower operating costs despite higher emissions levels than other techniques.

c) Plants could choose the appropriate technique according to their own needs. However, legal monitoring data is limited, and it still needs further research and verification as for long-term stability.

d) While technical upgrading, ceramic tile plants also need to improve operation and management measures, improve technical operation capabilities, and make the treatment of air pollutant emissions more effective.

IOP Conf. Series: Earth and Environmental Science **728** (2021) 012010 doi:10.1088/1755-1315/728/1/012010

7. References

[1] Li, H.-D.; Fang, Z.-B. Research on the Development of Chinese Ceramic Tile Industry under One Belt and One Road Initiative. *Journal of Hebei University (Philosophy and Social Science)*. 2018, 43, 126-137.

[2] European Commission. Integrated Pollution Prevention and Control (IPPC). Reference Document on Best Available Techniques in the Ceramic Manufacturing Industry. *Spain: European IPPC Bureau*, 2007, 1-260.

[3] Ministry of Ecology and Environment. *Technical guide for pollution prevention and control in the ceramic industry*: HJ2304-2018. Beijing: China Environmental Publishing Group, 2018.

[4] The State Department. *Planning for Ecological Environment Protection in the 13th five-year Plan*. The State Department: Beijing, China, 2016. http://www.gov.cn/xinwen/2016-12/05/content_5143464.htm

[5] Ministry of Ecology and Environment. *Compiling instruction of technical guide for pollution prevention and control in ceramic products manufacturing industry (Draft for Comments)*. Ministry of Ecology and Environment: Beijing, China, 2018. http://www.mee.gov.cn/gkml/sthjbgw/stbgth/201808/W020180828602523700152.pdf

[6] European Union. *Council Directive 96/61/EC of 24 September 1996 concerning integrated pollution prevention and control*. The Council of The European Union, 1996.

[7] Official Journal of the European Union. *Directive 2010/75/EC on industrial emissions (integrated pollution prevention and control)*. Official Journal of the European Union: Aberdeen, 2010.

[8] US EPA. U.S. *Clean Air Act, Section 111*. US EPA: Washington DC, 1990.

[9] US EPA. U.S. *Clean Air Act, Section 112*. US EPA: Washington DC, 1990.

[10] Liu, X.-Y.; Wen, Z.-G. Best available techniques and pollution control: a case study on China's thermal power industry. *Journal of Cleaner Production*, 2012, 23, 113-121.

[11] Wang, B.; Sun, Q.-H.; Hu, X.-W.; Bai, L.; Fang, L.; Lin, X.-J.; Yang, X.-S.; Wang, J. Screening of Best Available Techniques for Lead Smelting Pollution. Prevention and Control. *Journal of Environmental Engineering Technology*, 2011, 1, 526-532.

[12] Yang, L.-Y.; Yu, H.-B.; Wang, Q.-S. Research on the best available technology of Cleaner production in VC industry. Environmental Protection and Circular Economy, 2012, 35-39.

[13] Liu, X.-J.; Chang, L.-C.; Lin X.-J.; Zhu, M. Wang, S.-L. Assessment of best available technologies for dyeing and finishing of the textile industry. Cleaner Production, Energy-Saving and Emission Reduction, 2014, 5, 153-157.

[14] Ministry of Environmental Protection. Discharge standard of pollutants in the ceramic industry: GB 25464-2010. China Environmental Science Publishing House: Beijing, China, 2010.

[15] Ministry of Environmental Protection. Discharge standard of pollutants in ceramic industry（GB 25464-2010）Modification order. Ministry of Environmental Protection: Beijing, China, 2014.

Acknowledgments

This work was supported by 2018 Beijing Colleges and Universities High-level Talents Cross-training "Practical Training Program" Project, National Key Research and Development Plan Key Special Projects of China (No.2016YFC0208100).

IOP Conf. Series: Earth and Environmental Science **728** (2021) 012011 doi:10.1088/1755-1315/728/1/012011

Evaluation of Chromium Application in the Steel Industry in China: Implications on Environmental Quality

Baba Imoro Musah[1*], Lai Peng[1], Yifeng Xu[1]

1 Department of Environmental Science and Engineering, School of Resources and Environmental Engineering, Wuhan University of Technology, Luoshi Road No. 122, Hongshan, Wuhan, Hubei, 430070, PR China.
E-mail: babamusah94@yahoo.com

Abstract. This study provides vital analysis of environmental consequences of the steel industry in China on outdoor air quality, surface, and groundwater quality as well as global production of Chromite ore, its associated cost on health locally and internationally. The steel industry emissions currently amount to 50, 200, and 300 μg/m^3 for $PM_{2.5}$, SO_2, and NO_2 respectively exceeding WHO recommended emissions values. Out of 365 cities in China, 70% of these cities representing 256 cities in 2018 had their annual mean particulate matter emission being 44.1 μg/m^3 which is significantly higher than the WHO standard. Imported chromite from 1996 to 2016 amounted to 106,090,000 Mt by China. The production of crude steel increased from 870.9 to 928.3 Mt in 2017 and 2018 respectively. The proportion of good groundwater quality reduced from 26.9 to 25.4% in three years. Nearly 68% of total chromium emissions come from the metallurgical industry with an annual increment of 15.6% with an annual average $PM_{2.5}$ value of 53 ug/m^3 which is significantly very high.

1. Introduction

China is endowed with deposits of many mineral resources in minable quantities such as fossil fuels such as coal, natural gas, and oil. It is also known for being a top producer of aluminium, tungsten, zinc, copper, magnesium, lead, gold, mercury, iron, steel [1]. In 2016, China was the topmost country in the world with 24% percentage world mining whilst other countries such as the United States of America, Russia, and Australia recorded 11.5, 9.1, and 7.3% respectively. At the continental total mining outputs levels, Asia came top with 58.2% followed by North America 14.1%, Europe 8.5%, Oceania and Latin America followed with 6.9% each, and Africa coming behind with 5.4% [2]. Records showed that China is the number one producer of twenty-eight different raw mineral materials such as mineral fuels, industrial minerals, precious metal, non-ferrous metals, and iron and ferroalloy metals. This gives China a competitive edge and a very strategic position in terms of socio-economic growth at the same time serious challenges with environmental pollution. Apart from the 28 raw mineral deposits, China does not have sufficient chromium deposits despite that it's the top consumer of chromium and producer of steel in the world. Chromite resources in China are very limited and between 2011 to 2016 production had been on the decline. Globally, chromite resources that are economically viable in terms of quantities are estimated to be more than 12 billion metric tonnes. Interestingly, about 95% of the world's chromite deposits are geographically located in two countries namely South Africa and Kazakhstan [3]. Chrome ore is a mineral made up of chromium and iron oxide ($FeCr_2O_4$) but known commercially as chromite. Chromite is an essential material in the production of stainless steel given the fact that it serves as an alloying agent. Chromite is a very important and suitable material in the production of stainless steel given its anti-corrosion resistance properties [4]. The aims of this study were to *i) provide empirical chromite import and production analysis ii) examine the impact of the growing steel industry in China on environmental health and iii) analyze the contribution of the steel industry to national emissions and identify any associated health risks involved.* Given the increasing patronage and wide usage of stainless-steel products globally, and because China is the largest producer of stainless-steel and related products in the world, it is significant to do a critical assessment of its impact on the environment. The objectives were achieved through critical analysis of data obtained. It is anticipated that the results would help in understanding the intricacies involved in the chromite industry. To ensure exhaustive investigation, this study analyzed chromite production and imports data from 1996 to 2016. In addition, recent emissions and effluents discharge reports have been examined to ascertain the current trends. The implications of rapid growth of the steel industry in China on health and environmental sustainability. Effluents discharge and emissions of toxic gases have been analyzed critically.

2. Global chromite production

Chromite ore production in 2016 was 30.4 million metric tonnes (Mt) compared with a revised production quantity of 31.2 Mt in 2015, an indication of a 3% decline over the previous year. South African maintained its lead in chromite production in 2016 with 48% of worldwide production. On the contrary, global ferrochrome production went up to 10.9 Mt compared to 10.3 Mt in 2015. South Africa was the leading producer of ferrochrome accounting for 33% of world total production [5]. In 2017, the four largest chromium producers globally were South Africa (15 Mt), Kazakhstan (5.4 Mt), India (3.2 Mt), and Turkey (2.8 Mt) [6]. The International Stainless Steel Forum (ISSF) 2015 disclosed that South Africa has an approximated 72% of world chromite ore reserves followed by Zimbabwe with an estimated 12% of world chromite ore deposits [7]. Based on critical analysis of data, the total quantity of chromite produced by each of the world's top four countries such as Turkey, India, Kazakhstan, and South Africa in the last 21 years stood at 32,603,051; 56,490,042; 66,224,478, and 190,778,601 Mt respectively. The total quantity of chromite ore produced by the top four global producers in the past two decades amounted to 346,096,172 Mt [8-11]. This represented a significant quantum of global chromite ore produced in the past 2 decades. Fig. 1a and 1b showed the proportion of chromite ore produced by each of the top four global producers percentage-wise.

2.1 Dynamics of chromium production and trade

Chromium has wide applications in different industries globally ranging from metallic ceramics, paints and dyes, chrome plating, tannery and leatherwork, electric furnace, anti-corrosion resistance. Given the diverse applications of chromium by industries worldwide, Grand View Research Inc. (GVR), market research and consultancy firm in the US has predicted that global trade in chromium is expected to hit a 2.7% compound annual growth rate (CAGR) representing the market size of US$ 16.55 billion by the year 2025 from US$13.07 billion in 2016 [11]. The global chromium market growth is driven by rising demands by industries that manufacture aircraft and stainless steel as end-use. Additionally, the robust growth of the automobile and the electronic industry in the Asia Pacific region where China and India are located contribute significantly to the rising demand for chromium. China and South Africa play a very critical role in the production of ferrochrome for global market supply. In 2014, the total global production of ferrochrome by the two countries of 11,236 tonnes (t), this exceeded the global demand by +219 t. In 2015 and 2016, the total ferrochrome produced was 10,699 and 10,806 t respectively. This means that market demands for ferrochrome globally outweighed total production by China and South Africa by -141 and -851 t in 2015 and 2016 respectively. China imported a total of 106,090,000 (Mt) of chromite between 1996 to 2016 valued at US$23,821,101 based on analysis of import data. During this period, the lowest and highest chromite imports by China were 12,090,000 (Mt) in 2013 and 760,000 (Mt) in 1996 as vividly shown in Fig. 1c. Again, Fig. 1e showed ferrochrome production by China and South Africa from 2014 to 2016, total demand, supply, and projections regarding production between 2017 to 2021. Fig. 1f showed aggregated chromite ore production (%) by the top 4 countries from 2006 to 2016.

IOP Conf. Series: Earth and Environmental Science **728** (2021) 012011 doi:10.1088/1755-1315/728/1/012011

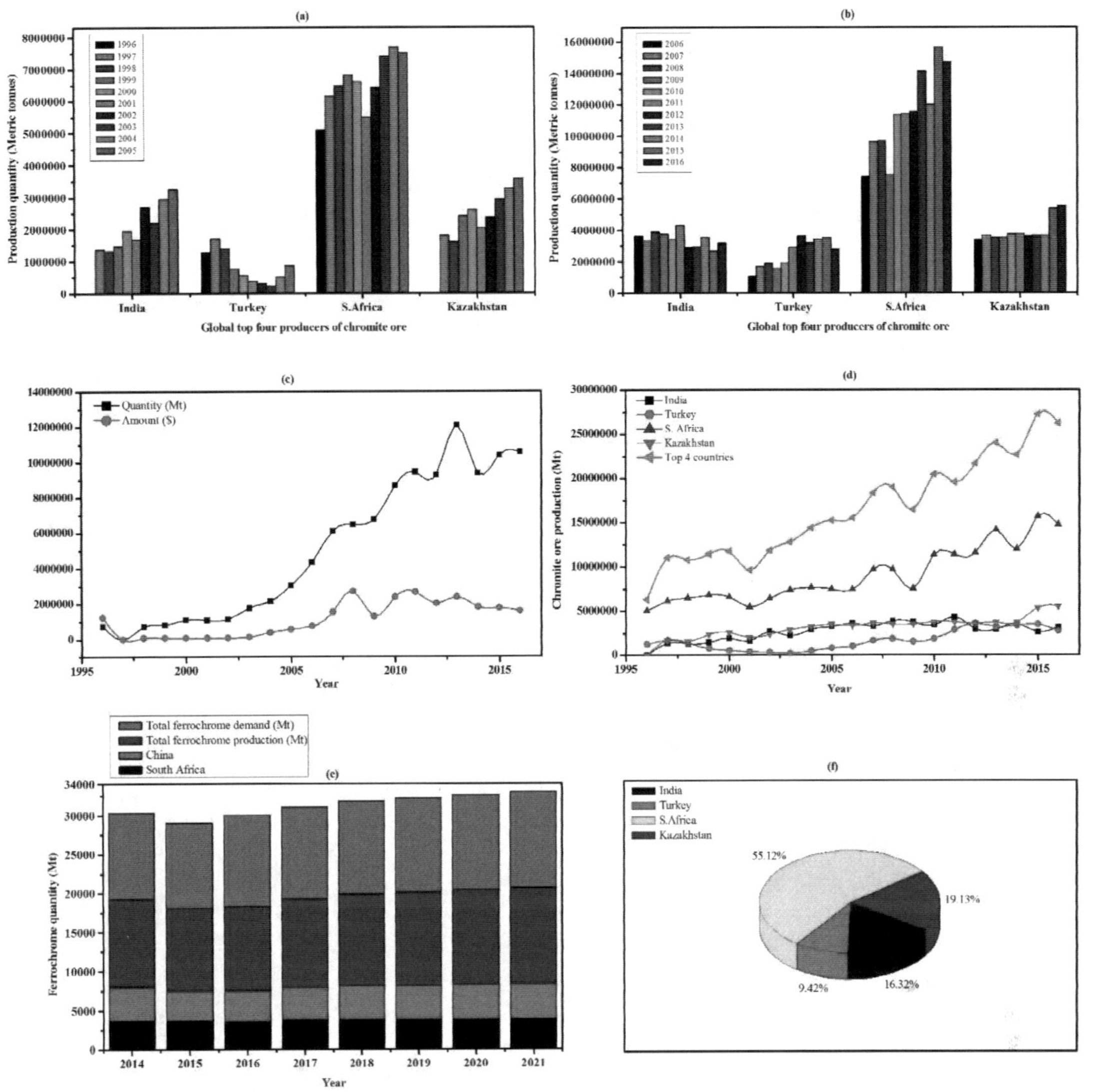

Authors' analysis (Data sources: US Geological Survey & Macquarie Group Ltd)

Figure 1. Global top 4 chromite ore producers (a) 1996 to 2005, (b) 2006 to 2016, (c) trends of China's chromium imports and costs from 1996 to 2016 (d) chromite ore production trends by top 4 countries from 1996 to 2016, (e) ferrochrome production by China and S. Africa, total global demand and supply and (f) aggregated chromite ore production (%) by the top 4 countries from 2006 to 2016.

3. Impact of the steel industry on environmental sustainability

Ferrochrome and steel industries require huge amounts of energy as well as the minerals for production to meet the ever-increasing demands for stainless steel and ferrochrome products. This also means that significant quantities of wastes are generated in the production processes. These wastes end up in the environment via disposal routes. These wastes contain substantial quantities of toxic pollutants, wastewater, residues, and other by-products that are generated in the course of the production. It has been reported that ferrochrome and steel industries emit dust with significant amounts of certain organic compounds and heavy metals. This is attributed to the use of electric arc furnaces in the steel manufacturing plants. Carbon dioxide (CO_2) is one of the constituents emitted into the environment in the production process. The International Iron and Steel Institute (IISI) has indicated that electric arc furnace plants emit nearly 0.7 tonnes of CO_2 per each tonne of steel produced. Similarly, integrated plants emit between 1.6 to 2.4 tonnes of CO_2 per tonne of steel produced [12]. The steel industry in

China discharged 1.73 Mt of sulfur dioxide (SO_2) in 2007 which represents 7% of the total national sulfur dioxide discharge. The same industry discharged 1.07 Mt of dust in the same year which represented 15% of total national dust emitted. The steel industry also discharged 228.5 Mt of solid wastes in 2007 which represented 13% of total national solid wastes discharged in that year. The quantity of wastewater discharged by the steel industry represented 8% of total national wastewater discharge [12]. China has been the world's number producer of stainless steel for over a decade now. In 2017, China has ranked 1st in world crude steel production with 870.9 Mt. Japan was 2nd with a production record of 104.7 Mt. India was 3rd with 101.5 Mt while the US ranked 4th with 81.6 Mt. The rest were Russia 5th position with crude steel production of 71.5 Mt, South Korea 6th position with 71.0 Mt, Germany 7th with a production record of 43.3 Mt. The last three countries were Turkey 8th, Brazil 9th and Italy 10th position with 37.5 Mt, 34.5 Mt and 24.1 Mt respectively. In 2018, China still maintained its dominant lead in the production of crude steel as the world's number one with 928.3 Mt, an increase in production by 57.4 Mt compared to 2017. India however displaced Japan to occupy the 2nd position in 2018 after it produced 106.5 Mt of crude steel. Japan dropped to 3rd position with a production record of 104.3 Mt. The US maintained its 4th position after its production in 2018 increased slightly to 86.6 Mt from 81.6 Mt in 2017. South Korea displaced Russia to occupy the 5th position with total crude steel production of 72.5 Mt. Russia came 6th in crude steel production of 71.7 Mt, but Germany also maintained its previous rank of 7th with 42.4 Mt despite a little decline in production. The remaining three countries all maintained their 2017 positions such as Turkey 8th with 37.3 Mt, Brazil 9th with 34.9 Mt, and Italy 10th with 24.5 Mt of crude steel [13].

3.1 Cost of pollution on economies

According to the Organization for Economic and Co-operation and Development (OECD) 2016 report, an estimated 3 million premature deaths in 2010 globally due to exposure to polluted outdoor air quality. It is further predicted that, if stringent measures are not put in place by countries worldwide, the number of premature deaths will increase to 6-9 million by 2060 annually. The cost of healthcare due to air pollution-related illness is predicted to increase from US$21 billion in 2015 keeping the 2010 US$ exchange rate constant to US$176 billion in 2060. Within the same period, loss of productivity as a result of the annual loss of working days is estimated to cost US$3.7 billion globally [14]. In 2018, the estimated cost of global air pollution due to exposure to PM2.5, NO_2, and O_3 about fossil fuel were US$2.7 trillion (3.1% of GDP), US$750 billion (0.9% of GDP), and US$524 billion (0.60% of GDP) respectively. It is predicted that 1.8 million people in Mainland China die prematurely as a result of exposure to outdoor air pollution yearly [15]. A total amount of 9.95 billion yuan equivalent to US$1.42 billion was spent as cost of air pollution control facilities operations such as desulfurization by the steel industry e.g smelting, metal products processing, and manufacture in China in 2016. In the same year, a total of 4.40 billion yuan an equivalent of US$629 million was expended by the steel industry in China as an operational cost for wastewater treatment facilities regarding smelting and ferrous metals processing. Nearly 423 Mt of raw steel was produced by the steel industry in China in 2006. Of this total production, the cost of operating air pollution control equipment stood at 23.5 yuan equivalent to US$3.36 per tonne of steel produced while the cost of operating wastewater treatment facilities amounted to 10.4 yuan equivalent to US$1.49 per tonne of steel produced in 2006. This excluded maintenance and other related operational costs [16].

3.2 Consequences of pollution on health

In 2019, South China Morning Post revealed that the Chinese government disclosed that curbing pollution was one of the "three critical battles" China continues to fight against apart from the eradication of poverty and financial risk control. Coal-powered plants and the steel industry are the biggest conduits of pollutant emissions in China. The steel industry in 2017 emitted 2.81 Mt of particulate matter (PM) which accounted for 20% of China's total PM emissions [17]. Additionally, China's steel industry generated 1.72 Mt of nitrogen dioxide (NO_2) and 1.06 Mt of sulfur dioxide. This means that current emission levels by the steel industry stand at 50 micrograms per cubic meter ($\mu g/m^3$) for $PM_{2.5}$, 200 $\mu g/m^3$ for SO_2, and 300 $\mu g/m^3$ for NO_2 per annum instead of WHO standard permissible emission limits of 10 $\mu g/m^3$, 20 $\mu g/m^3$ and 40 $\mu g/m^3$ for $PM_{2.5}$, SO_2 and NO_2 respectively [18]. It is anticipated that this year 2020, 60% of the steel manufacturing plants in areas in China will meet these standard benchmark values with the hope that by the year 2025, 80% of the steel manufacturing companies in China will meet the required standards. Between now and the next 5 years, China will continue to battle with toxic emissions into the atmosphere from the steel industry. Chinese government as part of its measures to encourage the steel industry in the country to meet these standards, companies that finish the upgrade by the stipulated time would be supported financially, given tax incentives, and protected by environmental policies.

China's rapid industrial and economic growth in the last 40 years came with associated consequences in terms of environmental pollution and human health. According to the World Bank, many cities in China are polluted due to emissions from industries. China is the largest contributor to Greenhouse Gases (GHG) globally. In 2016, China was the third country to launch a satellite to monitor CO_2 emission after the US and Japan [19]. In 2014, the annual average $PM_{2.5}$ concentration was 93 micrograms per cubic meter (ug/m^3), which was more than twice China's national standard limit for $PM_{2.5}$ of 35 ug/m^3 and the World Health Organization (WHO) permissible limit of 10 ug/m^3 for $PM_{2.5}$ and PM_{10} of 20 ug/m^3. Despite the efforts of China's government to curb air pollution, the World Bank's latest annual average $PM_{2.5}$ value is 53 ug/m^3 in 2017 which is still higher than both the national standard and WHO annual mean of 10 ug/m^3 [20]. The economic and social cost of China's air pollution cannot be underestimated considering the number of fatalities recorded yearly due to poor outdoor air quality. In a recent study on morbidity, mortality, and risk factors in China in 2019, it was revealed that the second and third highest cause of deaths in China in 2017 was *ischaemic heart disease and lung cancer* respectively with stroke coming first [21]. Zhou et. al, (2016) revealed that poor outdoor air quality was the 5th leading cause of premature deaths in China in 2013, thus contributing to 916,000 premature deaths in China in the same year.

In 2018, Green Peace in its air pollution control report card for 2017 revealed that 365 cities in China had a mean $PM_{2.5}$ concentration of 44.1 ug/m^3 as compared to 46.2 ug/m^3 in the previous year. Even though in comparison, 2017 mean $PM_{2.5}$ concentration represented a 4.5% drop but the average value of 44.1 ug/m^3 was still higher than the national standard permissible limit of 35 ug/m^3 and more than three times of WHO permissible mean $PM_{2.5}$ concentration of 10 ug/m^3. Similarly, 74 crucial cities across China had their mean $PM_{2.5}$ concentration dropping to 48 ug/m^3 in 2017 from 2013 mean value of 72 ug/m^3 which represented a decline of 33.3% in four years since China launched its pollution control campaign. Again, even though this looks progressive, the mean $PM_{2.5}$ concentration for those 74 key cities is almost five times higher than WHO permissible annual mean $PM_{2.5}$ concentration and also still higher than China's national standard limit. This is an indication that, despite the progress made, more work still needs to be done to achieve at least the national standard for $PM_{2.5}$. It is important to note that, based on the 2017 air pollution control scorecard assessment on 365 cities across China by Green Peace Initiative (GPI), 256 which represents over 70% of the total cities have $PM_{2.5}$ concentration higher than the national secondary standard of 35 ug/m^3. In terms of Ozone (O_3) concentration in the 365 cities surveyed across China, the mean concentration was 144.5 ug/m^{3}representing a 6.3% increment whilst NO_2 mean concentration was 30.5 ug/m$_3$ representing 3.3%. similarly, the 74 crucial cities had their mean O_3 concentration of 155 ug/m^3 representing an 11.5% increment over that of 2013 [15].

The Chinese government in February 2010 released data on its First National Census on Pollution Sources (FNCPS). The FNCPS identified a total of over 5.93 million sources of pollution in the country. Out of the total sources reported, 1.58 million of them are industrial sources dotted across the country. The total quantity of Chemical Oxygen Demand (COD) discharge was over 30.3 Mt out of which industrial sources COD discharge amounted to 5.6 Mt, the rest came from agricultural, landfills (leachate) and domestic sources [22]. This is an indication of the complex nature of the industrial effluents given the vast number of sources involved. A similar worrying outcome notice in the report was the amount of heavy metals effluents generated by industries. Heavy metals such as chromium, arsenic, lead, mercury, and cadmium wastewater reached 900 tonnes. It was revealed that SO_2 total emissions in the country stood at 23.20 Mt out of which over 21.20 Mt came from industrial sources while NO_x total emissions amounted to 17.98 Mt and 11.88 Mt out of the total emissions came from industrial sources. Solid and hazardous wastes were not left out in the 2010 FNCPS, because industrial solid and hazardous waste total discharge was 3.85 billion tonnes (Gt). Wang et al., (2016) reported that nearly 90% of total

PM emissions in the steel industry come from sinter plants and blast furnaces in China [23]. These statistics are frightening given the fact that between 2010 and 2020, industrial production especially in the steel industry continues to rise significantly. This makes the fight against environmental pollution from industrial sources more cumbersome and complicated than it can be imagined.

3.3 Impact of pollution on water resources

China has a good number of water resources from fresh to marine water bodies. However, due to increasing demand for water resources as a result of urbanization and increasing population growth, put pressure on the available water resources. More troubling is the fact that water quality deterioration continues to occur due to anthropogenic activities such as pollution. The pressure is brought to bear on groundwater for purposes of industrial and agricultural activities due to the contamination of surface water bodies. In 2007, the World Bank estimated that the overall cost of water scarcity and groundwater depletion which could be attributed to water pollution-related challenges was in the tune of 147 billion yuan [24]. The steel industry contributes significantly to chromium pollution in water bodies. An inventory study on chromium discharge into water bodies and atmospheric emissions in China by Cheng et al. (2014) revealed that approximately 13,400 tonnes of chromium were discharged into water bodies from six categories of industries in 20 years. The metallurgical industry alone accounted for nearly 68% of the total chromium emission with a yearly increment of 15.6% [25].

Table 1. Freshwater quality changes between 2013 to 2016

Name of River	**Year**	**Total water sections under National monitoring**	**Water quality status (%)**	**Change in water quality (%)**
			Grade I-III	**Grade I-III**
Songhua River Basin	2013	108	55.7	+4.5
	2016		60.2	
Huaihe River Basin	2013	180	59.6	-6.3
	2016		53.3	
Haihe River Basin	2013	161	39.1	-1.8
	2016		37.3	
Liaohe River Basin	2013	106	45.5	-0.2
	2016		45.3	
Yangtze River Basin	2013	510	89.4	-7.1
	2016		82.3	
Yellow River Basin	2013	137	58.1	+1
	2016		59.1	
Pearl River Basin	2013	165	94.4	-4.8
	2016		89.6	

Authors' analyzed results (Data source: Ministry of Environmental Protection, China)

Anthropogenic activities continue to pose severe threats to environmental quality. From Table 1, it can be observed that between 2013 to 2016, apart from Songhua River and Yellow River Basins which witnessed 4.5% and 1% improvement in the Grade I to III water quality over the period, the rest of the five fresh surface water bodies recorded some level of pollution. The percentage of Grade I to III water quality of the Huaihe River Basin with 180 water sections that are under national monitoring got polluted moderately by 6.3%. Similarly, Grade I to III water quality of the Haihe River Basin with 161 national monitoring sections, Grade I to III water quality of the Liaohe River Basin with 106 national monitoring sections, Grade I to III water quality of the Yangtze River Basin with 510 national monitoring sections and Grade I to III water quality of the Pearl River Basin with 165 national monitoring sections were polluted by 1.8, 0.2, 7.1 and 4.8% respectively compared to the 2013 water quality status of Grade I to III. According to the 2013 State of the Environment Report on China by the Ministry of Environmental Protection, out of a total of 4,778 identified sites (wells) for national groundwater quality monitoring in China, excellent water quality was 10.4% (497 sites), good water quality was 26.9% (1,285 sites), relatively good quality was 3.1% (148 sites), relatively poor quality stood at 43.9% (2, 098 sites) and severely poor quality was 15.7% (750 sites) correspondingly [26]. In 2016, the number of groundwater quality monitoring sites (wells) increased to 6, 124 as captured in the State of the Environment in China report. Out of the total number of groundwater monitoring sites by the Chinese government, 10.1% (619 sites) had excellent quality, 25.4% (1,555 sites) had good quality, 4.4% (269 sites) had relatively good quality, 45.4% (2,780 sites) had poor quality and 14.7% (900 sites) had very poor groundwater quality [27]. It can be observed that even though the total number of groundwater monitoring sites increased from 4,778 in 2013 to 6,124 in 2016, the percentage of excellent water quality reduced from 10.4% to 10.1% respectively. Within the same period, the percentage of total good groundwater quality also reduced from 26.9% in 2013 to 25.4% in 2016 respectively. The analysis showed that for both excellent and good quality groundwater sites witnessed some level of pollution between 2013 and 2016.

4. China's environmental monitoring and protection strategies

The launch of the CO_2 satellite by China to monitor emissions released into the atmosphere was a giant step. Similar monitoring of outdoor air quality takes place in 338 cities in China. These are important measures put in place by the Chinese government to mitigate the increasing air pollution.

Chinese government adoption of a "Master Plan for Eco-environmental Protection of Lakes with Fine water Quality 2013-2020". Under this master plan, the Chinese central government allocated 1.6 billion yuan to help protect 27 lakes in China with fine water quality. Additionally, 1.8 million environmental inspectors were recruited by the Chinese government to monitor regularly 0.81 million enterprises to ensure adherence and compliance with environmental standards and regulations. Nearly 10,000 environmental breaches, non-compliance, hazards, and risks were identified in the process. As part of measures to address the challenges of environmental pollution in the country, the Chinese government through the State Council introduced certain policy interventions aimed at curbing pollution. The 12th Five-Year Plan (2011-2015) is one of such policy interventions recently. The FYPs set out targets such as safe disposal and treatment of solid wastes, controlling the high occurrence of heavy metals pollution, strengthen environmental supervision system, enhance the development of legal regimes, increase investments all aimed at curbing environmental pollution.

The implementation of the 12th Five-Year Plan chalked some successes including a reduction in total COD discharge by 12.5% and a 14.3% reduction in SO_2 emissions in comparison to 2005 statistics [28]. However, the 13th Five-Year Plan on the protection of the environment released in 2016 showed that SO_2 and COD emissions were still high approximately 20 Mt [29]. Again, 78.4% of cities in China could not meet the standards set for outdoor air quality, 3.2% of the days were classified as extreme pollution. Moving forward, the Chinese government and all other relevant stakeholders ought to join forces together by possibly reviewing some of the existing interventions and tighten the loose ends. This would ensure that drastic measures are taken to fight against the increasing spate of outdoor air pollution.

5. Conclusion

The analysis revealed China's chromium consumption and importation trends. South Africa showed significantly dominates in chromite ore and ferrochrome production globally. Global ferrochrome production though intermittently recorded some slowdown, there has been increasing trends generally owing to increasing demand and application. The top four chromite-producing countries globally were South Africa, Kazakhstan, India, and Turkey respectively in 2017. China continues to lead the rest of the world in the production of steel and would likely continue to dominate in this industry for many years to come given its production advantages.

6. References

[1] Wang K 1944 Mineral Resources of China: With Special Reference to the Nonferrous Metals, *Am. Geogr. Soc.*, vol. **34**, no. 4, pp. 621–635.
[2] Government of Austria 2018 World mining data 2018: Iron and ferroalloy metals, non-ferrous metals, precious metals, industrial minerals, mineral fuels.
[3] US Geological Survey 2019 Mineral Commodity Summaries, February 2019.
[4] KPMG 2018 KPMG commodity insights bulletin – chromite, November, p. 8.
[5] US Geological Survey 2016 Minerals Yearbook [Chromium Advanced Release].
[6] Kay A 2018 4 Countries with the World's Highest Chromium Production, *Investingnews.com/*,2018.[Online].Available: https://investingnews.com/daily/resource-investing/industrial-metals-investing/chromium-investing/top-chromium-producers/. [Accessed: 17-Feb-2020].
[7] ISSF 2015 Stainless Steel in Figures.
[8] US Geological Survey 2005 Minerals Year Book [Chromium].
[9] US Geological Survey 2001 Minerals Year Book [Chromium].
[10] US Geological Survey 2010 Minerals Year Book [Chromium].
[11] Grand View Research Inc. 2018 Chromium Market Size Worth $16.55 Billion By 2025 |CAGR:2.7%.[Online].Available:https://www.prnewswire.com/news-releases/chromium-market-size-worth-1655-billion-by-2025--cagr-27-grand-view-research-inc-673764983.html. [Accessed: 17-Feb-2020].
[12] SEAISI 2008 Dealing with Environmental Pollution in the Iron and Steel Industry: The China Case Study, *South East Asia Iron and Steel Institute* [Online]. Available: http://www.seaisi.org/News/662/Dealing+with+Environmental+Pollution+in+the+Iron+and+Steel+Industry:+The+China+Case+Study. [Accessed: 26-Feb-2020].
[13] Worldsteel Association 2019 World Steel in Figures.
[14] OECD 2016 *The Economic Consequences of Outdoor Air Pollution*. Paris: OECD Publishing.
[15] Green Peace 2020 Toxic air: the price of fossil fuels, South East Asia.
[16] AAM 2009 An Assessment of Environmental Steel Industry in China, Washington.
[17] Xie E 2019 China targets the steel industry as the latest front in its war on pollution, *South China Morning Post*, Beijing.
[18] IEA Clean Coal Centre 2016 Emission standards and control of $PM_{2.5}$ from a coal-fired power plant, London.
[19] Agence France Presse 2016 China launches carbon dioxide monitoring satellite, *Phys.org*[Online].Available:https://phys.org/news/2016-12-china-carbon-dioxide-satellite.html. [Accessed: 28-Feb-2020].
[20] World Bank 2017 $PM_{2.5}$ air pollution, mean annual exposure (micrograms per cubic meter).[Online].Available:https://data.worldbank.org/indicator/EN.ATM.PM_{25}.MC.M3. [Accessed: 27-Feb-2020].
[21] Zhou M *et al.* 2019 Mortality, morbidity, and risk factors in China and its provinces, 1990 – 2017: a systematic analysis for the global burden of disease study 2017, *Lancet*, vol. **394,** no. 10204, pp. 1145–1158.
[22] Government of China 2010 1st National census on pollution sources, *Bulletin*. [Online]. Available:http://www.china.org.cn/china/2010-02/09/content_19394384.htm. [Accessed: 29-Feb-2020].
[23] Wang K, Tian H, Hua S, Zhu C, Gao J, Xue Y, Hao J, Wang Y and Zhou J 2016 A comprehensive emission inventory of multiple air pollutants from iron and steel industry in China : Temporal trends and spatial variation characteristics," *Sci. Total Environ.*, vol. **559**, no. 19, pp. 7–14.
[24] World Bank 2007 Cost of pollution in China: economic estimates of physical damages.
[25] Cheng H, Zhou T, Li Q, Lu L and Lin C 2014 Anthropogenic chromium emissions in China from 1990 to 2009, vol. **9,** no. 2.

[26] Ministry of Environmental Protection 2014 The 2013 report on the state of the environment in China, Beijing.
[27] Ministry of Environmental Protection 2017 The 2016 report on the state of the environment in China, Beijing.
[28] Ministry of Environmental Protection 2011 The 12th Five-Year Plan for environmental protection (2011-2015), Beijing.
[29] The 13th five-year plan for economic and social development of the People's Republic of China 2016–2020, Beijing.

Comparison and Analysis of Important Product Standards and Technical Indicator Systems in the Papermaking Field-Taking Facial Tissue as An Example

Jing Li[1], Wenlun Qiu[2] and Lifei Yang[3]

1 Associate professor, China National Institute of Standardization, China
2 Senior engineer, China National Pulp and Paper Research Institute Co., Ltd., China
3 Senior engineer, China National Institute of Standardization, China
E-mail: lijing123@cnis.ac.cn

Abstract. This article summarizes the current status of domestic and foreign standards for paper products and the standard system, compares and analyzes the domestic and international standards for facial tissue which is key product in the papermaking industry, combines the production technology, characteristics, and requirements of the product to establish key technologies index list for facial tissue, carries on comprehensive analysis of product standard index levels, testing methods, etc., and gives opinions and suggestions on product standard formulation and revision.

1. Introduction

Papermaking, a basic industry of raw materials, is one of the important pillar industries in national economy.[1] Paper and paperboard are main products of the papermaking industry, mainly including printing paper, cultural paper and office paper, packaging paper, technical paper, household paper.[2] As an important consumer product for daily use, paper products (such as facial tissue, toilet paper, paper diaper, sanitary pad, and copy paper) are closely concerned with people's daily life. With the continuously enhancement of people's standards of consumption and safety awareness, the consumption of paper products attracts more attention and shows an increasing trend year by year.

Since the implement of the Plan for Deepening the Reform of Standardization Work in 2015, China has successively released a series of important standard documents to put forward the standard comparison and analysis requirements for product quality improvement and brand building. The report of the 19th National Congress of the CPC pointed out that we need to deepen the supply-side structural reform and make improvement according to international standards. The Guiding Opinions of Implementing the Quality Improvement Action issued by the CPC Central Committee and the State Council proposed to "speed up the upgrading of standards and implement the enterprise standard forerunner system". The state has launched a special action for over ten thousand of enterprises from thousands of industries in hundreds of cities[3] and opinions on the implementation of enterprise standard "forerunner" system[4] for the purpose of actively carrying out the comparative evaluation of standards, giving full play to the fundamental, guiding and strategic role of standardization in quality improvement, improving product quality, and enhancing the quality of industrial development.

2. Domestic and foreign standards for paper products and the current status of the standards systems

2.1 Standards for Paper Products in China

After years of system construction, a standard system in which mandatory standards match recommended standards, national standards industry standards complement each other, and basic standards combine specific product standard at levels has been built for the paper products industry.

The National Paper Industry Standardization Technical Committee (TC141) is responsible for standardized technology of paper products in China. By the end of 2019, there were 468 existing standards in the paper industry, including 348 national standards and 120 industrial standards; by standard property: there were 1 mandatory standards, 467 recommended standards.

2.2 Standards for Paper Products in Foreign Countries

The following contents focus on the standard management systems, standardization mechanisms and standard system construction of the International Organization for Standardization (ISO), EU, Germany, the United Kingdom and Japan. [5] [6]

(1) ISO international standard

The ISO/TC6 "Technical Committee on Paper, Paperboard and Pulp" is an ISO organization responsible for developing relevant international standards concerning test methods and quality specifications of paper, paperboard and pulp, and its subordinate paper and paperboard technical sub-committees (ISO/TC6/SC2) are responsible for the preparation and revision of relevant test methods and standards for paper and paperboard. The ISO/TC6 is responsible for the standards of the general test method and pulp test method. The ISO/TC6 standard system consists of three parts: standards for basic general method, standards for paper and paperboard test method, and standards for pulp test method. At present, ISO/TC6 has a total of 177 papermaking standards, including 2 technical guidance documents. The standards formulated by ISO/TC6 mainly focus on testing methods. Most standards for papermaking test methods in China are formulated in compliance with ISO standards.

(2) European Union (EU)

The EU standard management system is composed of EU technical regulations, new approach directives, EU coordinative criteria and EU conformity assessment procedures. EU directives are about "basic requirements", namely, basic health and safety requirements that must be met before the product is put on the market. The task of the European standardization institution, on the other hand, is to develop corresponding technical specifications (namely the "coordinative criteria") that meet the basic requirements of the directives. When a product complies with these technical specifications, it can be determined that the product meets the basic requirements of the directives. According to EU regulations, all products covered by the new approach directives related to safety, hygiene, health and environmental protection must go through the corresponding conformity assessment procedures to get a CE mark before entering the EU market. A product with a CE mark on it indicates that the product complies with the basic safety, hygiene, health or environmental protection requirements stipulated in EU new approach directives and can be bought or sold freely in the EU market. At present, the relevant directives and resolutions involved include: REACH Regulations, and Eco-Label Requirements of EU Directive for Absorbent Hygiene Products (2014/763/EC) and Tissue paper (2009/568/EC).

(3) The United Kingdom (UK)

In the UK's standard management system, the Standard and Technical Regulation Division of the Department of Trade and Industry is the competent department in charge of standard establishment, test and certification of policies at the policy level. Specific standard, test, and certification management functions are performed by two government agencies: The standard management function is performed by the British Standards Institution (BSI), and the test and certification management functions are performed by the United Kingdom Accreditation Service. BSI is managing 240,000 existing British standards and 2,500 professional standards committees, with more than 230,000 standard committee members. In addition, thousands of standard projects are being developed by BSI each year. In terms of the standards for children's paper, the BSI mainly focuses on test method standards, with a small number of product standards. ISO standards are directly taken by countries as their national test method standards in most cases.

(4) Germany

Germany implements a non-governmental standardized management system authorized by the government. The government does not directly interfere in the operation of the country's overall economy. Deutsche Industrie-Norm (DIN) is a non-profit private organization founded in 1917. According to the agreement signed with the federal government of Germany, the DIN is a standardization competent authority which represents Germany in international and European standardization activities and manages domestic standardization activities, including the formulation and revision of standards. DIN standards are voluntary standards. When formulating standards, relevant interested parties shall determine whether to formulate relevant standards according to the market demand, with enterprises that master advanced technology playing the dominant role. Based on extensive participation by scientific research institutions, governments, individuals, the formulation of standards shall adhere to the principle of openness and transparency. Most standard committee members should be enterprises, and they shall put forward the need for developing standards and bear most of the cost derived from research and standard formulation. In German, all standards, technical regulations, technical regulations, technical rules, technical procedures, and technical specifications are collectively referred to as technical specification documents, and they are divided into three levels: DIN standards, technical rules, and technical regulations. German DIN Group owns German Conformity Assessment Company, DQS, European Certification Company and other famous certification organizations. In the field of children's paper, similar to the UK's standards, German standards also focus on test methods. The test method standards released by ISO or EN are directly taken as German standards.

(5) Japan

In Japan's standard management system, the Ministry of Economy, Trade and Industry (former Ministry of International Trade and Industry) is responsible for the formulation, revision, promulgation of industrial standardization regulations in addition to related administrative management. The specific work is carried out by the Japan Industrial Standards Committee (JISC). The draft of Japanese industrial standards (JIS) is proposed by the competent minister or non-governmental organizations and reviewed and published by the JISC. JIS standards developed accordingly are Japan's national standards. The Ministry of Health, Labor and Welfare (MHLW) is a ministry of the Japanese government. In the field of health, the Ministry of Health, Labor and Welfare of Japan has equivalent functions of some departments in China, including the National Health Commission, China Food and Drug Administration, National Development and Reform Commission (medical services and drug price control), Ministry of Human Resources and Social Security (medical insurance), Ministry of Civil Affairs (medical assistance), General Administration of Quality Supervision, Inspection and Quarantine (frontier sanitary quarantine). The Ministry of Health, Labor and Welfare of Japan including the Health Administration and Medical & Food Administration, is responsible for the national administration of food, drugs, cosmetics, medical devices and biological preparations, including the formulation of standards for these products.

(6) The United States (US)

With American National Standard Institute (ANSI) as the coordinating body, the American standards system consists of three parts: the national standards system, federal government standards system, and professional standards system of professional groups. In the United States, it is considered that mandatory standards may inhibit the improvement of productivity. Standards that are cited by law and formulated by government departments, however, are generally mandatory standards.

The standards developed by the Technical Association of the Pulp and Paper Industry (TAPPI) have become general standards recognized globally in the paper industry. TAPPI standards involves six parts: fibrous material and pulp test, paper and paperboard test, non-fibrous material test, packaging container test, building material test, and test procedures. Currently, there are 234 standards. The association has developed standards only for test methods other than product standards.

3. Comparative analysis of the standards for facial tissues and paper products at home and abroad

3.1 Domestic and Foreign Standards

Facial tissues are a kind of disposable sanitary paper made from plant fibers and processed by slitting and folding, and they are water-absorbing, soft, clean and convenient to use. Facial tissues mainly

include: paper tissues, paper towels, and paper napkins. They are widely used in restaurants, at tables, in families or used as carry-on personal belongings. As facial tissues have become one of the necessities in people's life, the quality and standard of facial tissues attract extensive attention. [7][8]
In the comparison of facial tissue standards, a total of 4 national standards in Chinese mainland and Taiwan Province are selected, involving EU's REACH regulations and facial tissue standards of other 4 countries (the United Kingdom, the United States, Canada, and South Africa). Table 1 shows 4 facial tissue standards in Chinese mainland and Taiwan Province. Table 2 shows 7 facial tissue regulations and standards in foreign countries.

Table 1. Facial tissue standards in China

No.	Standard, Regulation Number and Title	Remarks
1	GB/T 20808-2011 *Facial Tissue*	National standard
2	GB 15979-2002 *Hygienic Standard for Disposable Sanitary Products*	National standard
3	CNS 2386-2014 *Paper Napkin*	Standard in Taiwan Province
4	CNS 4150-2014 *Facial Paper*	Standard in Taiwan Province

Table 2. Facial tissue standards in foreign countries

No.	Standard, Regulation Number and Title	Remarks
1	*REACH Regulations*	EU
2	BS 1439-1992 *Specification for Single-Ply Wet-Creped Paper Toweling*	The UK
3	DOD MIL-T-37968-1986 *Tissue (for Dentistry)*	the US
4	CGSB-9.4-94 *Facial Paper for Public Use*	Canada
5	SANS 1887-3: 2008 *Facial Paper*	South Africa
6	SANS 1887-4: 2008 *Tissue*	South Africa
7	SANS 1887-7: 2008 *Paper Napkin*	South Africa

3.2 Comparative Analysis on Safety Indicators of Facial tissue Standards in China and Foreign Countries

3.2.1 Safety Requirements in Domestic Facial Tissue Standards. For Chinese facial tissue standards, the safety indicators mainly refer to raw materials, chemical safety and hygiene safety:
(1) Raw materials
GB/T 20808-2011 *Facial Tissue* clearly stipulates that making facial tissues from toxic and hazardous materials is prohibited. Instead, virgin fibrous materials such as wood, grass, bamboo should be used. Making facial tissues from any recycled paper, printed matter, paper product and other recycled fibrous material is prohibited. The application of deinking agent is prohibited.
The current CNS 2386-2014 *Paper Napkin* (Taiwan) stipulates that 100% virgin chemical pulp or pulp mixed with part of virgin mechanical wood pulp shall be used as raw materials; the current CNS 4150-2014 *Facial Paper* stipulates that the raw materials shall be 100% virgin pulp.
(2) Chemical safety
In order to ensure the product performance or meet the need for process control during the production of facial tissues, decolorizer may be used in the pulping process. Some companies even add fluorescent brightener to make tissue whiter. To cope with this problem, GB/T 20808-2011 stipulates that the detection of migrating fluorescent brightener is not acceptable. Moreover, GB/T 20808-2011 *Facial Tissue* has regulated the ash contents: wood fiber ≤1.0%, wood fiber ≤4.0%. The control over ash indicators can directly restrain the adding of a large amount of materials, which avoids the shedding of

IOP Conf. Series: Earth and Environmental Science **728** (2021) 012012 doi:10.1088/1755-1315/728/1/012012

powder and indirectly limits the use of secondary fibers in facial tissues.
When making facial tissues, we need to add some chemicals such as softening agent, wet-strength agent. Therefore, chemical residues in finished paper may pose a risk for safe use. However, there is no existing specified standard for acrylamide.
CNS 2386-2014 *Paper Napkin* (Taiwan) and CNS 4150-2014 *Facial Paper* (Taiwan) stipulate that products shall not contain migrating fluorescent substances, and the chemicals, pigments and dyes used shall comply with relevant health regulations.
(3) Hygiene safety
GB/T 20808-2011 *Facial Tissue* also stipulate that hygienic indexes of facial tissue (aerobic bacterial count, colon bacillus, pseudomonas aeruginosa, staphylococcus aureus, hemolytic streptococcus, fungal count) shall meet the requirements of GB 15979-2002 *Hygienic Standards for Disposable Sanitary Products*. According to the requirements, the aerobic bacterial count in facial tissue products shall be less than and equal to 200 cfu/g, the fungal count shall be less than and equal to 200 cfu/g, and coliform bacteria and pathogenic pyogenic bacteria (pseudomonas aeruginosa, staphylococcus aureus and hemolytic streptococcus) shall not be detected.

3.2.2 Safety Requirements in Foreign Facial Tissue Standards. EU: Safety indicators related to facial tissue in REACH Regulations include: The content concentration of acrylamide shall be less than and equal to 0.1%; the mass fraction of nonyl phenol and nonylphenol polyoxyethylene ether shall be less than and equal to 0.1%.
The UK: BS 1439-1992 *Specification for Single-Ply Wet-Creped Paper Toweling* stipulates that wet crepe wipes should be free of fluff or fiber, and the pH value of dye residues in the hands shall range from 4 to 10 after use.
The US: DOD MIL-T-37968-1986 *Tissue (for Dentistry)* stipulates that facial tissues shall be made from 100% bleached wood pulp.
Canada: CGSB-9.4-94 *Facial Paper for Public Use* stipulates that tissue paper can be made from virgin pulp or recycled pulp.
South Africa: SANS 1887-3: 2008 *Facial Paper* stipulates that facial paper shall be made from processed cellulosic fiber with a pH value ranging from 4.4 to 10.0; SANS 1887-4: 2008 *Tissue* stipulates that tissue paper shall have a pH value ranging from 4.4 to 10.0; SANS 1887-7: 2008 *Paper Napkin* stipulates that paper napkins shall be made from processed cellulosic fiber with a pH value ranging from 4.4 to 10.0.

3.2.3 Comparison Analysis of Facial Tissue Safety Indicators. In terms of raw materials, China's national standard GB/T 20808-2011 Facial Tissue, CNS 2386-2014 Paper Napkin (Taiwan) and CNS 4150-2014 Facial Paper (Taiwan), the US's DOD MIL-T-37968-1986 Tissue (for Dentistry) have strict requirements for raw materials, that is, 100% virgin fibrous materials shall be used as raw materials. Canada's CGSB-9.4-94 Facial Paper for Public Use stipulates that facial paper can be made from virgin pulp or recycled pulp. China has a stricter requirement for raw materials than that of Canada.
In terms of chemical safety, China's facial tissue standard stipulates that the detection of migrating fluorescent brighteners and the application of deinking agent are prohibited. The standards of Taiwan stipulate that the product shall not contain migrating fluorescent brighteners, and the chemicals, pigments and dyes used shall comply with relevant health regulations. The UK's facial tissue standard stipulates that the dye shall not remain in the hands when facial tissue is used. South Africa's standard stipulates the pH range of facial tissue. The EU's *REACH Regulations* stipulate the limits of acrylamide, nonylphenol and nonylphenol polyoxyethylene ether, while the limits of these substances have not been included in China's standard at present.
In terms of hygienic safety, China's GB/T 20808-2011 *Facial Tissue* stipulates that hygienic indicators of facial tissue shall meet the requirements of GB 15979-2002 *Hygienic Standard for Disposable Sanitary Products*. In other countries, hygienic indexes are not included in standards for facial tissues.

3.3 Comparison and Analysis of the Indicators for Physical Properties of Facial Tissue in Domestic and International Standards

3.3.1 Indicators for Physical Properties of Facial Tissue in Domestic Standards. The indicators for physical properties in the GB/T 20808-2011 Facial Tissue in China mainly include brightness, tensile index (CD), wet tensile strength (MD), capillary rise (CD), and softness. Table 3 gives the details.

Table 3. Major indicators for physical properties in GB/T 20808-2011 *Facial Tissue*

<table>
<tr><th colspan="2" rowspan="3">Indicator</th><th rowspan="3">Unit</th><th colspan="3">Provision</th><th rowspan="3">Testing method</th></tr>
<tr><th colspan="2">Superior</th><th rowspan="2">Qualified</th></tr>
<tr><th>Supersoft</th><th>Ordinary</th></tr>
<tr><td colspan="2">Brightness</td><td>%</td><td colspan="3">≤90.0</td><td>GB/T 7974</td></tr>
<tr><td colspan="2">Tensile index (CD)</td><td>N.m/g</td><td>≥1.0</td><td>≥2.10</td><td>≥1.50</td><td>GB/T 12914-2008</td></tr>
<tr><td colspan="2">Wet tensile strength (MD)</td><td>N/m</td><td>≥10.0</td><td>≥14.0</td><td>≥10.0</td><td>GB/T 12914-2008</td></tr>
<tr><td rowspan="2">Capillary rise (CD)</td><td>Single-ply</td><td rowspan="2">mm/100s</td><td colspan="2">≥20</td><td>≥15</td><td rowspan="2">GB/T 461.1</td></tr>
<tr><td>Double-ply or multi-ply</td><td colspan="2">≥40</td><td>≥30</td></tr>
<tr><td rowspan="2">Average softness of MD and CD</td><td>Single-ply or double-ply</td><td rowspan="2">mN</td><td>≤40</td><td>≤85</td><td>≤160</td><td rowspan="2">GB/T 8942</td></tr>
<tr><td>Multi-ply</td><td>≤80</td><td>≤150</td><td>≤220</td></tr>
</table>

According to CNS 2386-2014 *Paper Napkin* (Taiwan): size tolerance: ±5%, color be: ≥75%, and quantification (single-ply): ≤30g/m^2. According to CNS 4150-2014 *Facial Tissue* (Taiwan): size tolerance: ±5%, quantification (single-ply): more than 12.0g/m^2, tensile strength (CD, double-ply): ≥80gf/25mm, wet tensile strength (MD, double-ply): ≥23.5N/m, and water absorbency in the first minute (single-ply): ≥12mm.

3.3.2 Indicators for Physical Properties of Facial Tissue in International Standards. The relevant standards of the UK, South Africa, Canada and the US have provisions on the indicators for physical properties of facial tissue:

(1) The quality indicators specified in the BS 1439-1992 *Specification for Single-ply Wet-creped Paper Toweling* include: size (the area of each individual towel shall be not less than 645cm^2, and in no case shall the dimension of the shorter edge of each sheet of toweling be less than 20.3cm); wet tensile strength: ≥1.8N/15mm; rate of absorbency: ≤30s/0.02ml.

(2) According to the DOD MIL-T-37968 *Towel, Paper (Dental)* in the US: bursting strength of towel (12-ply): ≥15points, quantification (500-ply, 24×36 IN): ≥10.5tbs, water absorbency: ≤10s, brightness: 80%, and wet tensile strength (CD ≥0.5kg, MD ≥0.7kg).

(3) The CGSB-9.4-94 *Facial Tissue for Institutional Use* in Canada has provisions on dry tensile strength (average of MD and CD for single-ply ≥40N/m), wet tensile strength (average of MD and CD for single-ply ≥6N/m), and water absorbency (≤20s).

(4) The indicators specified in the SANS 1887-3: 2008 *Facial Tissue* in South Africa mainly include: quantification, dry tensile strength (average of MD and CD): ≥93N/m, softness: ≤120mN, and brightness: ≥78%.

The indicators specified in the SANS 1887-4:2008 *Tissue* mainly include: quantification, water absorption rate: ≤50s, total water absorption (single-ply ≥180%, double-ply ≥250%), dry tensile strength (MD ≥350N/m, CD ≥160 N/m), and softness: ≤300mN.

The indicators specified in the SANS 1887-7:2008 *Paper Napkin* mainly include: quantification, water absorption rate (type 1 ≤75s; type 2 ≤10s; type 3 ≥300s), total water absorption (≥180%), dry tensile strength (type 1: CD ≥53N/m, average of MD and CD ≥107 N/m; type 2: CD ≥33 N/m, average of MD and CD ≥67N/m; type 3: CD ≥67N/m, average of MD and CD ≥120N/m), softness (type 1 ≤275mN; type 2 ≤145mN; type 3 ≥390mN), and brightness ≥68%.

*3.3.3 Comparison Analysis of the Indicators for Physical Properties of Facial Tissue in Domestic and International Standards.*The indicators for physical properties of facial tissue are mainly strength, water absorbency, and softening property. The strength, water absorbency and softening property of facial tissue are mainly compared and analyzed this time, as shown in Table 4, Table 5 and Table 6, respectively.

(1) Strength

Table 4. Comparison of strength of facial tissue

Country	Standard No. and Tile	Wet tensile strength or index	Dry tensile strength (bursting strength)
China	GB/T 20808-2011 *Facial Tissue*	Qualified ≥10.0N/m, Superior (ordinary) ≥14.0N/m, Superior (supersoft) ≥10.0N/m	Tensile index (CD) Qualified ≥1.50 N.m/g, Superior (ordinary) ≥2.10 N.m/g Superior (supersoft) 1.00N.m/g
Taiwan	CNS4150-2014 *Facial Tissue*	Wet tensile strength (MD, double-ply) ≥23.5 N/m	Tensile strength (CD, double-ply) ≥3.14 N/m
The UK	BS 1439-1992 *Specification for Single-ply Wet-creped Paper Toweling*	Wet tensile strength ≥12N/m	Unspecified
Canada	CGSB-9.4-94 *Facial Paper for Public Use*	Wet tensile strength (average of MD and CD for single-ply ≥6N/m)	Average of MD and CD for single-ply ≥40N/m
South Africa	SANS 1887-3: 2008 *Facial Paper*	Unspecified	Average of MD and CD 93N/m
	SANS 1887-4: 2008 *Tissue*	Unspecified	MD ≥350 N/m CD ≥160 N/m
	SANS 1887-7: 2008 *Paper Napkin*	Unspecified	Type 1: CD ≥53N/m, average of MD and CD ≥107 N/m, type 2: CD ≥33 N/m, average of MD and CD ≥67 N/m, type 3: CD ≥67 N/m, average of MD and CD ≥120 N/m
the US	DOD MIL-T-37968-1986 *Tissue (for Dentistry)*	CD ≥0.5kg Average of MD and CD ≥0.7kg	Bursting strength (12-ply ≥15points)

The wet tensile strength (MD) is an important indicator to reflect the wet strength of facial tissue. Because the facial tissue will contact with water or sweat when they are used, the tissue should have a certain wet strength, which can avoid the phenomenon of cracking, dry linting and paper dust & fluff when wet. According to the national standard of China, the wet tensile strength (MD) of the qualified for facial tissue shall be larger than and equal to 10.0N/m, that of the superior (ordinary) shall be larger than and equal to 14.0N/m, and that of the superior (supersoft) shall be larger than and equal to 10.0N/m; according to the Canadian standard, the average wet tensile strength of MD and CD (single-ply) shall be larger than and equal to 6N/m; according to the standard of the UK, the wet tensile strength shall be larger than and equal to 12N/m; according to the standard in Taiwan, the wet tensile strength (MD, double-ply) shall be larger than and equal to 23.5 N/m. According to the US standard, the wet tensile strength (CD) shall be larger than and equal to 0.5kg, and the average of MD and CD shall be larger than and equal to 0.7kg. On the whole, all countries have provisions on the indicator for wet tensile strength. Due to differences in testing methods, it is not easy to judge the indicator value, but the indicator level is basically the same.

Moreover, it is stipulated in the national standard of China, the tensile index (CD) of the qualified for facial tissue shall be larger than and equal to 1.50 N·m/g, that of the superior (ordinary) shall be larger than and equal to 2.10 N·m/g, and that of the superior (supersoft) shall be 1.00N·m/g; the standards in Taiwan, Canada and South Africa do not define the tensile index (CD), but instead specify the tensile strength (CD). The US standard provides for the bursting strength.

(2) Water Absorbency

Table 5. Comparison of water absorbency of facial tissue

Country	Standard No. and Tile	Capillary rise (CD)	Others
China	GB/T 20808-2011 *Facial Tissue*	Qualified: Single-ply ≥15mm/100s, double-ply or multi-ply ≥30mm/100s Superior: Single-ply ≥20mm/100s, double-ply or multi-ply ≥40mm/100s	Unspecified
The UK	BS 1439-1992 *Specification for Single-ply Wet-creped Paper Toweling*	Unspecified	Total absorbency ≥0.93g, rate of absorbency ≤30s/0.02ml
Canada	CGSB-9.4-94 *Facial Paper for Public Use*	Unspecified	Water absorbency ≤20s
South Africa	SANS 1887-4: 2008 *Tissue*	Unspecified	Water absorption rate (≤50s), total water absorption ≥180% for single-ply, ≥250% for double-ply
	SANS 1887-7: 2008 *Paper Napkin*	Unspecified	Total water absorption ≥180% Water absorption rate: type 1: ≤75s, type 2: ≤10s, type 3: ≥300s
the US	DOD MIL-T-37968-1986 *Tissue (for Dentistry)*	Unspecified	Water absorbency: ≤10s
Taiwan	CNS 4150-2014 *Facial Paper*	Water absorbency in the first minute (single-ply): ≥12mm	-

Capillary rise (CD) is to measure the capillary rise within 100s, which reflects the water absorbency of tissue. In the Chinese standards, the indicator for water absorbency of facial tissue is mainly capillary rise (CD). In international standards, water absorbency, total water absorption, water absorption rate, and other indicators are mainly specified. The standards in Taiwan stipulate the capillary rise in the first minute. Due to differences in testing methods and time, it is impossible to determine the level of the set indicator value.
(3) Softening Property

Table 6. Comparison of softening property

Country	Standard No. and Tile	Softness
China	GB/T 20808-2011 *Facial Tissue*	Qualified: Single-ply or double-ply products ≤160mN, multi-ply products ≤220mN Superior: (ordinary) Single-ply or double-ply products ≤85mN, multi-ply products ≤150mN (supersoft) Single-ply or double-ply products ≤40mN, multi-ply products ≤80mN
The UK	BS 1439-1992 *Specification for Single-ply Wet-creped Paper Toweling*	Unspecified
Canada	CGSB-9.4-94 *Facial Paper for Public Use*	Unspecified
South Africa	SANS 1887-3: 2008 *Facial Paper*	Softness ≤120mN
	SANS 1887-4: 2008 *Tissue*	Softness ≤300mN
	SANS 1887-7: 2008 *Paper Napkin*	Softness: type 1 ≤275 mN, type 2 ≤145 mN, type 3 ≥390mN
the US	DOD MIL-T-37968-1986 *Tissue (for Dentistry)*	Unspecified
Taiwan	CNS 2386-1987 *Paper Napkin*	Unspecified
	CNS 4150-2008 *Facial Tissue*	Unspecified

Softness is also an important technical indicator of facial tissue, which reflects the softening property of facial tissue. Softening property is the most direct feeling of consumers when using facial tissue. The softening property of facial tissue in the Chinese standard is mainly measured by the softness. Compared with the standard for softness in South Africa, its classification of facial tissue is more explicit, so the requirements for softness are relatively more complete. The standards for facial tissue in the UK, Canada, the US, Korea, and Taiwan have no provisions on the softness.
Furthermore, the standards in China for facial tissue and products thereof have also limited the brightness, while those in other countries have no provisions on the brightness.

4. Conclusion

Upon comparison and analysis of domestic and international standards for facial tissue, we get some conclusions.
In terms of the indicators for hygienic safety, the GB 15979-2002 *Hygienic Standard for Disposable Sanitary Products* is implemented for facial tissue in China, which is more stringent than that in other countries.
In terms of the provisions on chemical safety, according to the Chinese standard for facial tissue, no chemical substances such as deinking agents, fluorescent brighteners and the like can be used in the production process. However, the REACH regulations in the EU have limited the amount of acrylamide,

nonylphenol, and NPE. Therefore, the regulations on the addition of harmful chemical substances such as acrylamide need to be further completed in the standards for facial tissue in China.
In terms of raw materials, according to the Chinese standard, no toxic and harmful raw materials shall be used for facial tissue, but virgin fiber materials such as wood, grass, and bamboo shall be used, and raw materials cannot be recycled paper, paper prints, paper products, and other recycled fibrous materials. However, according to the Canadian standard, the raw materials can be virgin pulp or recycled pulp. Therefore, China has stricter requirements on raw materials of facial tissue.
In terms of physical properties, there are differences in indicator names, testing methods, and expression methods of results in the standards of different countries, which are not comparable. On the whole, they are basically the same. The Chinese standards have provisions on the grade and plies of the products, which are more detailed and complete.
The following suggestions are proposed:
(1) Efforts shall be made to complete the technical contents of the product standards in China based on the comparison analysis conclusions, such as supplementing acrylamide, heavy metals and other safety indicators in facial tissue, so as to further improve the applicability of the standards and product quality level, thus protecting the health and safety of consumers.
(2) The quality safety of raw materials is closely related to the finished products. Strengthening the formulation of the standard for raw materials can ensure the safety of finished products. In addition, efforts shall be made to study and develop the standards for testing method of harmful chemical substances and the indicators for product comfort in paper products, and improve the safety and comfort of domestic diaper products by completing standard methods.
(3) Production enterprises should strengthen the management and control over the life cycle including raw materials and production processfrom the source, establish a raw material acceptance and traceability system, improve process conditions, and eliminate product safety risks as much as possible[9][10]. They should benchmark against the most stringent standards actively to establish their own enterprise standards and improve the product quality with high standards.

5. References

[1] China Paper Association 2019, China Technical Association of Paper Industry. White Book on Sustainable Development of China's Paper Industry. *China Paper Newsletters*, **3**,10-19
[2] Cao Pufang, Chen Youqing 2016. Research on the Status Quo and Development Trend of China's Paper-marking Industry. *Journal of Development-Oriented Finance Research*, **3**,63-73
[3] Information on http://www.sac.gov.cn
[4] Information on http://www.sac.gov.cn
[5] Shi He, Liu Wen, Li Hongkai 2018. Current Situation and Suggestion on the Paper Industry Standards in China. China Standardization, **11**,108-111
[6] Guo Xian, Liu Jing, Xiao Chengxiang, Sun Tingting 2016. Comparison Analysis on the Domestic and International Standardization Organization Systems. *China Standardization*, **2**,51-57
[7] Chen Yong, Zou Yuhui, Zheng Yongwei 2012. Research on the Standards for Paper Towel and Measures to Improve Quality Safety. *Quality and Standardization*, **7**,34-37
[8] Guo Caiyun 2020. Overview of World Paper Industry in 2018. *China Pulp & Paper*, **3**,78-82
[9] Li Dong 2019. Discussion on Green Sustainable Development of China's Paper Industry. *Paper Science & Technology*, **6**,26-29
[10] Chen Chunxia, Wu Shubin, Cheng Hao, Wei Qi 2019. Discussion on the Situation of the Tissue Paper Production in Compliance with Good Manufacturing Practices. *China Pulp & Paper*, **10**,56-61

Acknowledgement
This research was financially supported by Science and Technology Planning Project of Administration of Quality Supervision, Inspection and Quarantine (Project No. 2017QK014), Science and Technology Planning Project of the State Administration for Market Regulation (Project No. 2019MK119) and the President Foundation of China National Institute of Standardization (Project No. 242019Y-6691).

Nursing Home Wind Environment Evaluation in Northern China - A Case Study in Dezhou

Xuwei Ni*

Tianjin University Research Institute of Architectural Design & Urban Planning, Tianjin, China
E-mail: nxw_tju@163.com

Abstract. As the rapid development of institution-based care for the elderly people, the nursing home's outdoor wind environment research has become an important topic including wind potential evaluation as well as wind comfort analysis. In this study, the wind environment of a nursing home located in Dezhou, northern China has been evaluated quantitatively according to CFD methods. The parameters related to the wind environment including average wind velocity, wind velocity variation range and area proportion of shadow wind, comfort wind and risky wind regions has been calculated. Results showed the annual average wind velocity of the nursing home's outdoor environment was 1.69m/s and its variation range was 40.41%. The area proportions of shadow wind, comfort wind and risky wind regions were 23.00%, 75.35% and 1.65%. This indicated that the wind environment optimization of the nursing home should be focused on decreasing the shadow wind regions and dismiss the risky wind regions, and relative optimization suggestions were proposed.

1. Introduction

The problem of the aging population in China is becoming serious due to the continuous increasing of the elderly people which has been reached 25.4million and account for 18.1% of the total population in the end of 2019[1]. Otherwise, with the increasing demand of the high quality care service proposed by the elderly people, the disadvantages of the traditional home-based care mode emerge gradually, and more and more elderly people tend to choose institution-based care mode because of better medical facilities, experienced nursing assistants and same age friends. As the outdoor environment of the nursing home is the main space for elderly people's fitness activities and social communications, the outdoor environment design directly affects the physical and mental health of the elderly people resulting in an increasing investigations related to the nursing home outdoor environment design in recent year [2-3]. The wind environment evaluation is an important component of the outdoor environment because it not only affects the excising and living environment comfort of elderly people but also the safety of the pedestrians especially near high-rise buildings as well as sustainable development of the green buildings[4-5]. In this paper, a nursing home located in Dezhou, northern China has been chosen as the research object, and the wind environment of the nursing home has been evaluated according to CFD methods in order to discover the existing problems of the wind environment. The wind velocity distribution of the nursing home's outdoor environment in each season were simulated, and the annual wind environment has been quantitatively evaluated by area proportions of the shadow wind, comfort wind and risky wind regions which were divided according to the wind velocity. Based on these results, the regions with poor wind environment were identified clearly and the optimization suggestions has been also proposed.

2. Modeling methodology

2.1. Model establishment

The aerial view of the nursing home is shown in Figure 1. The nursing home is adjacent to No.12 Wei Street in north and Meijiao Road in east covering an area of 69000m^2 with 300m in north-south direction and 230m in east-west direction. There are a public care center and a living service center planed in the northern side of the nursing home, and the elderly people's residential zone is planned in the southern side. There are five group apartments in the residential which are named Liangyi garden, Jinghe garden, Tingfang garden, Wenmei garden and Wangchun garden respectively with Chinese meaning of living a long and healthy life.

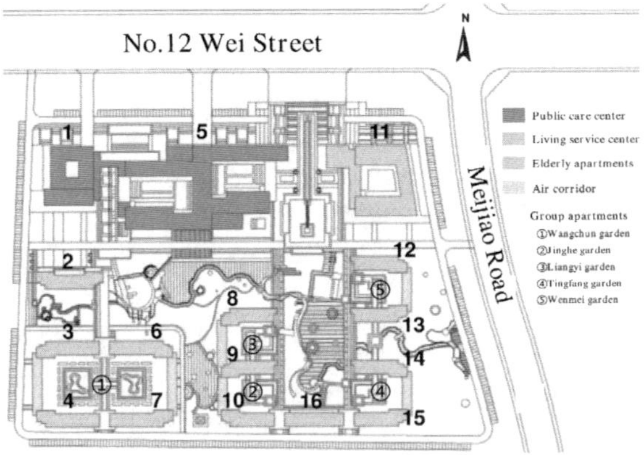

Figure 1. Aerial view of the nursing home

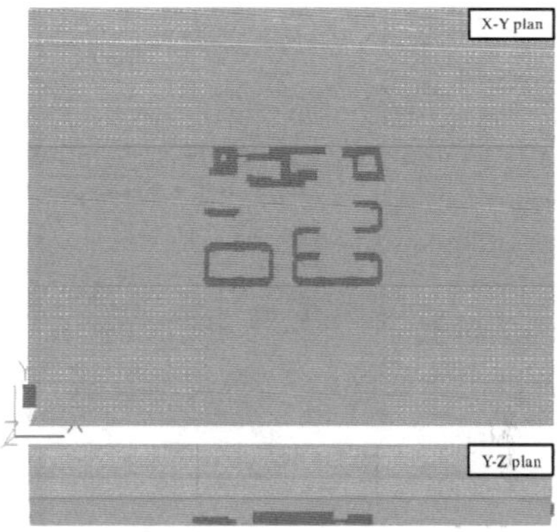

Figure 2. Simulation grids formation

The length, width and height of computational domain were all set as 3 times of the nursing home space, which was 900m, 690m and 132m respectively. The dense structured grids were used in the nursing home building area with an resolution value of 1m × 1m × 0.5m, and the sparse structured grids were used in other areas with an resolution value of 2m × 2m × 1m as shown in Figure 2. The total grids number reached 49,087,280, which was suitable to the present simulation.

2.2. Mathematical model

Considering both the accuracy and efficiency of the simulation, the standard k-ε turbulence model has been used to simulate the air flow of the outdoor environment, and the coefficients setting is shown in Table.1. Since the height of the computational domain was only 132m, which the air in this height could be considered as incompressible fluid, the boundary condition of the inlet was set as constant wind velocity. The wind velocity under different height (H) was calculated by $V= V_0(H/H_0)^n$ (where V_0 was the wind velocity of the reference height, reference height H_0 was set as 10m, and the roughness coefficient n was set as 0.16 as the recommendation value of suburbs[6]). Boundary condition of outlet was set as free outlet, and the boundary condition of the computational domain top was set as free sliding plane.

Table 1. Coefficients setting of standard k-ε turbulence model

C_μ	$C_{\varepsilon 1}$	$C_{\varepsilon 2}$	$C_{\varepsilon 3}$	$C_{\varepsilon 4}$	σ_k	σ_ε
0.09	1.44	1.92	-1.0	-0.33	1.0	1.22

3. Simulation conditions

3.1. Season division

The method of the season division in this paper was according to the related standard QX/T 152-2012[7]. If the average temperature was above or below 10°C for 5 consecutive days, the first day of this 5 consecutive days was defined as the start of spring or winter. If the average temperature was above or below 22°C for 5 consecutive days, the first day was defined as the start of summer or autumn. After analyzing the daily meteorological data of Dezhou in 2019, the season division was confirmed as shown in Table 2. According to the standard related to wind environment evaluation (GB/T 50378-2019[6]), the wind environment of spring and autumn are considered as one season for evaluating in this paper due to the close environmental temperature.

Table 2. Season division

Season	Spring	Summer	Autumn	Winter
Duration	Mar. 14-May 13	May 14-Sep. 14	Sep.15-Nov.16	Nov.16-Mar. 13

3.2. Time distribution characteristic of elderly people's outdoor activities
In order to accurately evaluate the wind environment of the nursing home, the time distribution characteristic of local elderly people's outdoor activities has been investigated according to the questionnaire and the results has been showed in Figure 3. It showed that the variation of time distribution of different seasons was limited, and was mainly concentrated in two time periods, one was from 6:00am to 12:00am, the other was from 2:00pm to 8:00pm. Therefore, only the wind environment of these two time periods (total 12 hours) in each season has been evaluated.

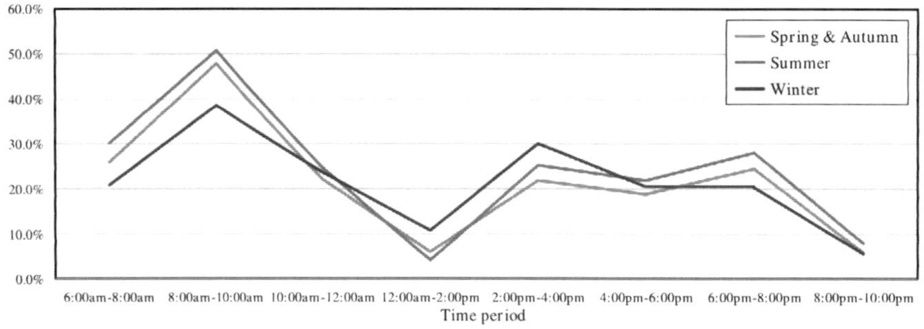

Figure 3. Time distribution characteristic of elderly people's outdoor activities

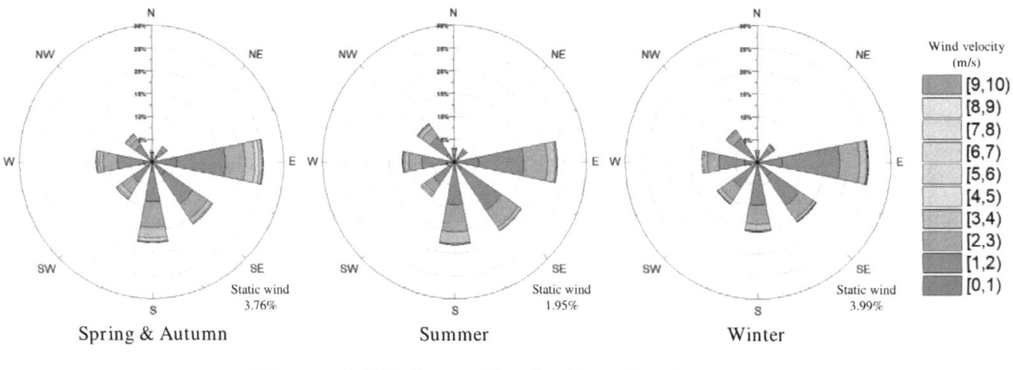

Figure 4. Wind rose distribution of each season

3.3. Simulation condition setting
According to the meteorological data, the wind rose distribution of each season has obtained as shown in Figure 4, and the static wind was defined as the wind velocity was below 0.1m/s. Results showed that both the wind direction and wind velocity distributions of each season were similar, the main wind

directions were all east, followed by south-east and south, and the main wind velocity ranged from 1m/s to 3m/s. The average wind velocity during elderly people outdoor activities (6:00 am -12:00am & 2:00pm - 8:00pm) of each wind direction in each season was listed in Table 3, and a total of 24 simulation conditions has been set.

Table 3. Wind direction and average velocity of each simulation condition

Wind direction	N	NE	E	SE	S	SW	W	NW
Spring and Autumn								
Simulation condition No.	1	2	3	4	5	6	7	8
Wind average velocity(m/s)	2.31	2.44	3.37	2.76	3.35	3.18	3.00	3.31
Proportion(%)	2.28	4.23	25.34	16.47	17.81	9.95	12.77	7.39
Summer								
Simulation condition No.	9	10	11	12	13	14	15	16
Wind average velocity(m/s)	1.83	2.38	2.80	2.67	3.17	2.75	3.17	2.94
Proportion(%)	3.09	3.56	23.45	18.28	18.21	9.27	11.96	10.22
Winter								
Simulation condition No.	17	18	19	20	21	22	23	24
Wind average velocity(m/s)	2.22	2.58	2.67	2.55	2.84	2.87	2.56	2.70
Proportion(%)	2.64	4.84	24.93	15.88	15.31	11.11	12.61	8.69

4. Results and discussion

In this part, the wind environment of the nursing home in each season has been discussed. The wind velocity distribution in the pedestrian height (1.5m) of each wind direction was first given. The average wind velocity and its variation range (standard deviation/average) were also calculated. Moreover, according to the wind velocity distribution, the outdoor environment has been divided into 3 types, which were shadow wind, comfort wind and risky wind regions. The shadow wind region was defined as the region where the wind velocity was lower than 1.0m/s, which against the diffusion of air pollutants and easily produced a stifling feeling to the elderly people, especially in summer. Considering the elderly people's physiological characteristics, cold and heat intolerant, the comfort wind region was defined as the region where the wind velocity was higher than 1.0m/s but lower than 3.3m/s[8]. The risky wind region was the region with wind velocity above 3.3m/s, the high velocity would cause respiratory and digestive diseases as well as cold and fever due to the rapid sweat evaporation, and also falling object hazards. All the area proportion of these three regions has been given, and the higher percentage of the comfort wind region and more narrow wind velocity variation range implied a better wind environment.

4.1. Wind environment evaluation in each season

The wind velocity distribution under different wind directions in spring and autumn is shown in Figure 5. It could be seen that the shadow wind region often appeared in public care center and living service center in the northern side of the nursing home, and also appeared in group apartments of Wangchun(zone ①), Jinghe(zone ②), and Tingfang gardens(zone ④). Besides, the distribution of the shadow wind region was much wider in the conditions with south-north direction wind than other conditions, this was because that all the buildings in the nursing home were south-facing, which defended and weakened the wind with south-north direction. The risky wind region often appeared in the passages with east-west direction, such as the passages between building No.2 and No.3, No.11 and No.12, No.13 and No.14, which was because of the acceleration effect when the wind flowed through adjacent tall buildings. The distribution of the risky wind region was wider in the conditions with east-west direction wind, but fortunately, the risky wind region was rare in the five group apartments zones.

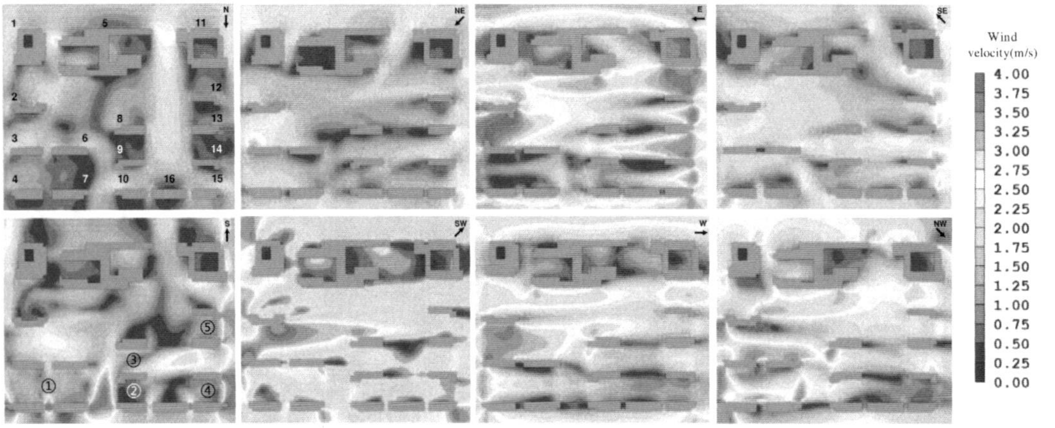

Figure 5. Wind velocity distribution in spring and autumn

The average wind velocity and its variation range, and the area proportions of shadow wind, comfort wind and risky wind regions are given in Table 4. The average wind velocity was 1.81m/s in spring and autumn, which was in a comfort level. The area proportions of shadow wind, comfort wind and risky wind region were 21.49%, 74.99% and 3.52%. In addition, the best wind environment was obtained in the conditions with southeast wind due to the minimum value of variation range and maximum proportion of comfort wind region. And the wind environment in the conditions with north and south wind were relatively poor.

Table 4. Wind environment related parameters in spring and autumn

	N	NE	E	SE	S	SW	W	NW	Weighted
Av. wind velocity(m/s)	1.23	1.51	2.17	1.73	1.58	1.85	2.02	2.08	1.81
Variation range(%)	43.40	37.42	42.10	33.14	55.60	44.44	38.70	37.82	40.76
Shadow wind region(%)	36.79	19.90	16.27	12.69	29.12	18.15	14.25	13.77	21.49
Comfort wind region(%)	63.21	80.10	76.79	87.31	66.33	77.92	83.57	82.37	74.99
Risky wind region(%)	0.00	0.00	6.95	0.00	4.55	3.93	2.18	3.83	3.52

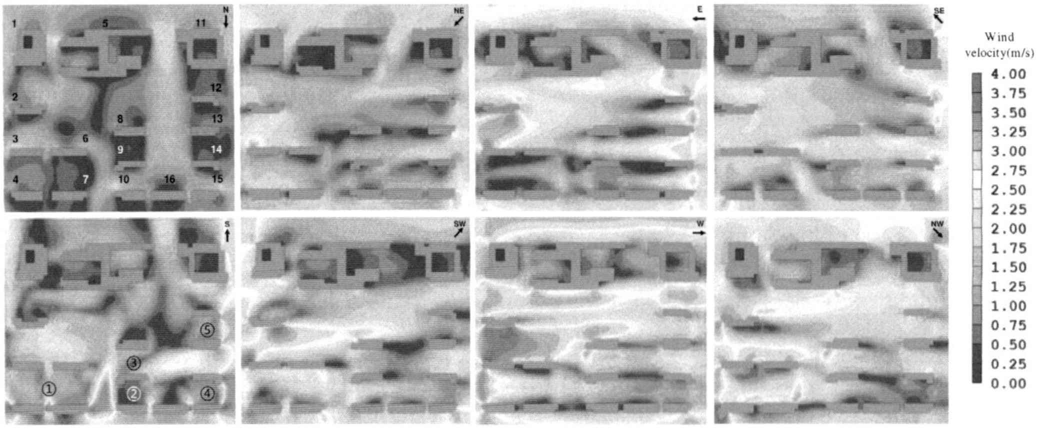

Figure 6. Wind velocity distribution in summer

The wind velocity distribution under different wind direction in summer is shown in Figure 6, which was similar to the results of spring and autumn. The calculation results of the wind environment related parameters are given in Table 5. The average wind velocity was 1.69m/s in summer, which was also in a comfort level. The area proportions of shadow wind, comfort wind and risky wind region were 22.46%, 76.40% and 1.13%. In addition, the best wind environment was obtained in the conditions with northeast wind. And the wind environment in the condition with north was the poorest.

Table 5. Wind environment related parameters in summer

	N	NE	E	SE	S	SW	W	NW	Weighted
Av. wind velocity(m/s)	0.98	1.68	1.81	1.68	1.50	1.61	2.13	1.86	1.69
Variation range(%)	42.30	33.10	41.87	33.38	54.73	44.08	38.87	37.29	40.92
Shadow wind region(%)	53.70	13.70	20.58	14.22	31.62	22.47	12.82	15.26	22.46
Comfort wind region(%)	46.30	86.30	79.12	85.78	65.68	76.79	83.64	83.93	76.40
Risky wind region(%)	0.00	0.00	0.29	0.00	2.69	0.75	3.54	0.78	1.13

The wind velocity distribution under different wind direction in winter is shown in Figure 7, which was similar to the results of spring, autumn and summer. The calculation results of the wind environment related parameters are given in Table 6. The average wind velocity was 1.56m/s in winter, which was also in a comfort level. The area proportions of shadow wind, comfort wind and risky wind region were 25.16%, 74.63% and 0.21%. In addition, the best wind environment was obtained in the conditions with southeast wind. And the wind environment in the conditions with north and south were the poorest.

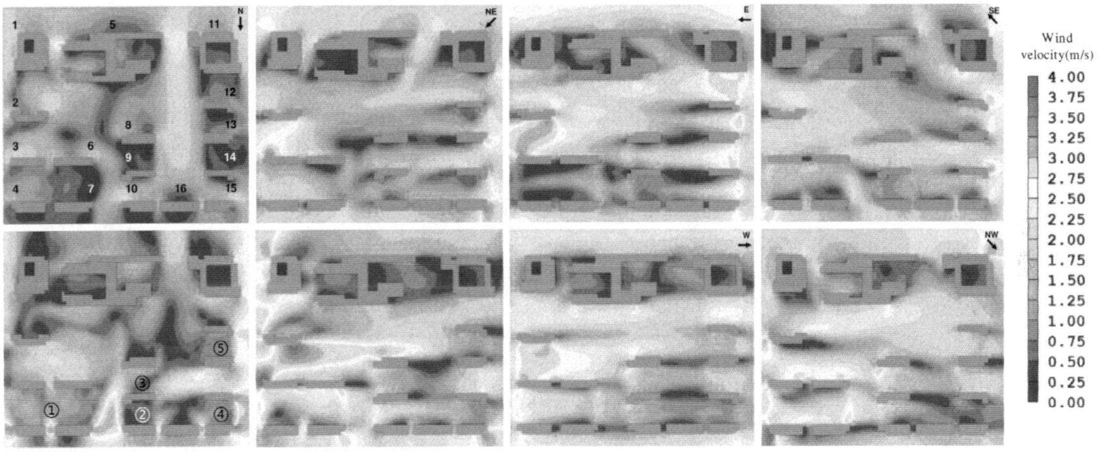

Figure 7. Wind velocity distribution in winter

Table 6. Wind environment related parameters in winter

	N	NE	E	SE	S	SW	W	NW	Weighted
Av. wind velocity(m/s)	1.18	1.59	1.74	1.61	1.37	1.68	1.76	1.71	1.56
Variation range(%)	43.40	37.76	41.23	33.14	53.61	44.23	36.62	37.16	39.49
Shadow wind region(%)	39.55	17.76	20.58	15.03	37.60	21.01	16.95	17.56	25.16
Comfort wind region(%)	60.45	82.14	79.38	84.97	62.31	77.82	83.05	81.85	74.63
Risky wind region(%)	0.00	0.10	0.03	0.00	0.10	1.17	0.00	0.58	0.21

4.2. Annual wind environment evaluation and optimization suggestions

The annual results of the wind environment related parameters are summarized in Table 7. It can be seen from the table that the average wind velocity in spring and autumn are higher than that in summer and winter, for all the seasons, the average wind velocities were in the comfort level with an annual value of 1.69m/s. The wind velocity variation range in each season was closed with an annual value of 40.41%. The area proportion of shadow wind region in winter was the highest which reached 25.16%, and the annual value was 23.00%. The area proportion of comfort wind region in summer was the highest which reached 76.40%, and the annual value was 75.35%. The area proportion of risky wind region in spring and autumn was the highest which reached 3.52%, and the annual value was 1.65%. Based on the analysis, the wind environment optimization of the nursing home should focus on the decrease of the shadow wind region area proportion. Besides, some typical risky wind regions should also be modified. The optimization of the shadow wind and risky regions would also narrow the wind velocity variation range simultaneously.

Table 7. Seasonal and annual wind environment related parameters

	Spring & Autumn	**Summer**	**Winter**	**Annual**
Av. wind velocity(m/s)	1.81	1.69	1.56	1.69
Variation range(%)	40.76	40.92	39.49	40.41
Shadow wind region(%)	21.49	22.46	25.16	23.00
Comfort wind region(%)	74.99	76.40	74.63	75.35
Risky wind region(%)	3.52	1.13	0.21	1.65

Figure 8 shows the main wind shadow and wind risky regions based on the wind velocity distribution of each season. It could seen that the wind shadow region was mainly appeared in the public care center and living service center. Because of the mechanical ventilation would be adopted to solve the ventilation problem of the patios in building No.1 and No.11, the main optimization should be focused on the structure of building 5. The suggestion was changing building 5 structure from "I" to "T", or using the space corridor to connect these two parallel buildings in order to avoid the formation of the approximative enclosure space. For the optimization of the wind shadow region between buildings No.8 and No.13, buildings No.9 and No.14, the suggestion was using out of place layout style rather than symmetrical layout style. For the wind shadow region appeared in southern side of the group apartments of zone ①, zone ② and zone ④, the suggestion was enlarging the gap between buildings No.4 and No.7, buildings No.10 and No.16, and buildings No.16 and No.15 in order to increase the air mass flow.

Figure 8. Main wind shadow and wind risky regions

The risky wind region often appeared in the passages with east-west direction, such as the passages between building No.2 and No.3, No.11 and No.12, No.13 and No.14. the optimization suggestions included planting evergreen arbors in the eastern and western side of nursing home to defend the wind with east-west direction, enlarging or dismissing the space between two parallel buildings (buildings No.2, No.3, No.11, No.12, No.13 and No.14) in order to weaken the acceleration effect caused by wind flowing through adjacent tall buildings.

5. Conclusion

In this paper, the wind environment of a nursing home located in Dezhou has been evaluated according to CFD methods in order to discover the existing problems of the wind environment. The wind distribution of the nursing home's outdoor environment in each season was simulated, and the annual wind environment has been quantitatively evaluated by the average wind velocity and its variation range, moreover, the area proportions of the shadow wind, comfort wind and risky wind regions were also calculated. Based on these results, the conclusions can be reached as the following.

1. The annual average wind velocity of the nursing home's outdoor environment was 1.69m/s and its variation range was 40.41%. the area proportions of shadow wind, comfort wind and risky wind regions were 23.00%, 75.35% and 1.65%.
2. The wind environment related parameters of different seasons were closed and the best wind environment was obtained in summer with the highest comfort wind region area proportion (76.40%). The area proportion of shadow wind region in winter was the highest which reached 25.16%, and the area proportion of risky wind region in spring and autumn was the highest which reached 3.52%.
3. The wind environment optimization of the nursing home should focus on the decrease of the shadow wind region area proportion and the suggestions included changing structure of building 5, using out of place layout style rather than symmetrical layout style of buildings No.8, No.13, No.9 and No.14, and enlarging the gap between buildings No.4 and No.7, buildings No.10 and No.16.
4. Besides, some typical risky wind region should also be modified, and the suggestions included planting evergreen arbors in the eastern and western side of nursing home, enlarging or dismissing the space between two parallel buildings (buildings No.2, No.3, No.11, No.12, No.13 and No.14).

6. References

[1] https://baijiahao.baidu.com/s?id=1656119076785202941&wfr=spider&for=pc (in Chinese)

[2] Othman A R, Fadzil F 2015 Influence of Outdoor Space to the Elderly Wellbeing in a Typical Care Centre *Procedia - Social and Behavioral Sciences* 320-329.

[3] Li J, Liu N 2020 The perception, optimization strategies and prospects of outdoor thermal comfort in China: A review *Building and Environment* **170**: 106614.

[4] Yung, E. H., Wang, S., & Chau, C. K. 2019 Thermal perceptions of the elderly, use patterns and satisfaction with open space *Landscape and Urban Planning* **185**: 44-60.

[5] Van Druenen T, Van Hooff T, Montazeri H, et al. 2019 CFD evaluation of building geometry modifications to reduce pedestrian-level wind speed *Building and Environment* **163**: 106293.

[6] Ministry of Housing and Urban-Rural Development of China 2019 Assessment standard for green building (GB/T 50378-2019) (in Chinese)

[7] China Meteorological Administration 2012 Devision of climatic season (QX/T 152-2012) (in Chinese)

[8] Zhang J, Li B, Zhang J, et al. 2014 Research and Practice of out-door environment optimal design fot apartments of the elderly based on sustainable design *Energy saving in building* **2**: 41-43.

Acknowledgments

This study is supported by Dezhou Hemujia nursing home project of Tianjin university research institute of architectural design & urban planning.

IOP Conf. Series: Earth and Environmental Science **728** (2021) 012014 doi:10.1088/1755-1315/728/1/012014

Effects of Variable Responses to Climate Change on the Interactions of Migratory and Resident Birds in Europe

Yuren Cao

Asociación Escuelas Lincoln, B1637 AOS La Lucila, Argentina
E-mail: ycao@lincoln.edu.ar

Abstract. Long-distance migrants and residents may respond to large-scale climate change by advancing egg laying dates. It is expected that variation in changes in avian laying dates among similar species affects their interactions in an ecological community. By analyzing long-term trends of the mean laying dates (MLD) of pied flycatcher *Ficedula hypoleuca*, collared flycatcher *Ficedula albicollis*, and great tits *Parus major* across 25 study sites in Europe from five published studies, the paper uses two-tailed t tests to show that variable adjustments in laying dates in different species have different implications on their interactions across Europe. It is found that pied flycatcher advanced laying dates significantly faster compared to great tits in Central Europe, gaining advantage in their competition for nest holes. Additionally, the study shows that pied flycatcher and collared flycatcher potentially share smaller overlap in laying dates in their sympatric breeding ground in Northern Europe, which may decrease the occurrences of hybridization, whereas the trend reverses in Central Europe. The results highlight that geographical variations in phenological responses to climate change have complicated effects on interspecific interaction, a novel research field that lacks empirical results. The conclusions of this study provide potential directions for empirical studies of climate change in the future.

1. Introduction

Birds have responded to large-scale climate change through changes in avian life-history events (phenology), particularly in the timing of migration and egg laying [1], although the extent of such adjustments may differ significantly within distinct geographical areas for a particular species [2]. The migratory pied flycatcher *Ficedula hypoleuca* has been reported to advance their arrival dates at breeding grounds in Europe in order to track the warming temperature [3, 4], yet limited adaptive ability to advance their phenology sufficiently causes a mismatch with local conditions such as food availability and may have profound consequences on interspecific interactions within biological communities, leading to declines in population sizes [5, 6]. Therefore, understanding variation between similar species in their responses to a changing environment is critical if we are to understand how climate change impacts the structure of biological communities that include both migratory birds and residents.

Although shifts in spring phenology of long-distance migrants such as the pied flycatcher have been well studied [7], it remains an open question as to how different responses between migrants and residents affect their interactions [8]. Recent evidence has indicated that resident birds have higher phenological sensibility than migratory birds among European cavity breeders, yet it is still unclear how this alters competitive interactions [9]. There are very few studies, to my knowledge, that have investigated the effects of variable responses to climate change on interspecific interactions between migrants and residents over different geographical areas.

The pied flycatcher, European great tits *Parus major*, and collared flycatcher *Ficedula albicollis* are ideal subjects for studying the effects of climate change because extensive long-term studies on their laying dates have been conducted at their breeding grounds in Europe and their interactions in an ecological community are relatively well-known. Here, this study uses mean laying dates across different sites in Europe as a proxy for the egg laying phenology of the studied species, which is obtained from five previously published studies [2, 10-13]. Given these mostly complete time-series of mean laying dates, two-tailed *t* tests are utilized to analyze the differences in long-term responses of laying dates with respect to geographical region and species. Finally, the implications of differential responses on the interspecific interactions of these species are discussed.

2. Methods

2.1. Study species and area

Pied flycatchers are long-distance migratory birds that overwinter in Western Africa and disperse throughout Europe during the breeding season. They occupy nest boxes on breeding grounds, allowing long-term studies to be conducted regarding their laying dates. Great tits are residents that generally breed earlier than pied flycatchers and actively compete with them for nest boxes, leading to possible influence in on breeding numbers and reproductive success [14]. Collared flycatchers are migratory birds closely related to the pied flycatcher. Although the two flycatchers are allopatric on most of their breeding grounds, there are overlapping ranges in central Europe and on the Gotland island in the Baltic, where they hybridize regularly [15].

Mean laying date (MLD) is defined as the average date of all first clutches that females in a population produce in a given year. In order to avoid reporting bias, all past studies that contain accurate long-term trends of mean laying dates of either of these three species of birds and that were available to the author were incorporated in this study. In total, long-term data from 24 study sites, as shown in Table 1, were used. The value of *n* refers to the number of years in the time-series for which the laying dates are obtained. The mean MLD is the mean of laying dates in the first five study years. The statistics of MLD-year, as well as the slope, are obtained from linear regression analysis. Experimental methods in these study sites differ, but in most instances, nest boxes are checked weekly in order to determine the laying dates. Since the changes of mean laying dates per year are of particular interest in this study instead of the actual laying dates, different experimental methods do not influence the results significantly as long as they are correct and consistent in each study site.

The study sites were subdivided into three areas: Western Europe includes the Netherlands and Belgium; Central Europe includes southern Germany, Switzerland, Czech Republic, and Hungary; Northern Europe is the region above the Baltic, whose latitude is approximately above 54°N. Firstly, it has been established that pied flycatchers take different migratory routes to each area. For example, a light geolocator study on pied flycatcher individuals confirms that populations in Netherlands and Belgium are originated from the western part of the species' distribution in Africa [16], whereas populations in central Europe may be originated in the eastern part in Africa [17]. Since migratory routes to each study sites are likely to be different according to their geographical locations, categorizing them into three regions helps reveal clearer, and potentially different, patterns of changes in laying dates. Secondly, since pied flycatchers and collared flycatchers are only sympatric in Central Europe and Gotland, it is only reasonable to use study sites in these regions in order to find variations in their responses to climate change.

Table 1. Long-term data from 24 study sites, obtained from five previous studies [2, 10-13]

study site	full name	coordinates	time period	n	mean MLD	slope MLD verses year	stats MLD-year
			Collared flycatchers *Ficedula albicollis*				
LH	Lanžhot	48° 40' N 16° 56' E	1961-2007	—	—	-0.20	$CI = 99.9\%$
LN	Lednice	48° 48' N 16° 46' E	1961-2007	—	—	-0.19	$CI = 99.9\%$
VN	Vranovice	48° 56' N 16° 35' E	1961-2007	—	—	-0.19	$CI = 99.9\%$
HK	Horka	49° 39' N 17° 11' E	1961-2007	—	—	-0.16	$CI = 99.9\%$
MR	Moravia	49° 50' N 17° 15' E	1973-2002	—	—	-0.193	$p = 0.024$
PM	Pilis mountains	47° 10' N 19° 09' E	1983-2002	16	6 May	-0.207	$p = 0.13$
GL	Gotland	57° 10' N 18° 20' E	1981-2002	22	26 May	-0.284	$p = 0.096$
			Pied flycatchers *Ficedula hypoleuca*				
HA	Harz	51° 53' N 10° 37' E	1980-2002	—	17 May	-0.397	$p < 0.001$
LR	Lahr	48° 17' N 7° 17' E	1981-2002	2	—	-0.53	$CI = 95\%$
HT	Harthausen	49° 17' N 8° 24' E	1973-2018	46	—	-0.31	$CI = 95\%$
GT	Goteborg	57° 43' N 11° 58' E	1980-2000	17	25 May	-0.157	$p = 0.064$
GN	Gunnebo	57° 40' N 12° 05' E	1980-1998	19	22 May	0.249	$p = 0.15$
BL	Borlange	60° 23' N 15° 30' E	1981-1999	19	26 May	0.128	$p = 0.41$
BA	Baulmes	46° 47' N 06° 31' E	1980-2002	23	18 May	-0.312	$p = 0.009$
BK	Buunderkamp	52° 01' N 05° 45' E	1984-2002	17	12 May	-0.358	$p = 0.006$
DW	Deelerwoud	52° 05' N 05° 55' E	1980-2002	22	11 May	-0.342	$p = 0.004$
HV	Hoge Veluwe	52° 02' N 05° 51' E	1980-2002	23	14 May	-0.466	$p < 0.001$
SP	Staphorst	52° 37' N 06° 17' E	1980-2002	23	12 May	-0.417	$p < 0.001$
WB	Warnsborn	52° 00' N 05° 51' E	1980-2002	23	11 May	-0.287	$p = 0.007$
LG	Lingen	52° 27' N 07° 15' E	1980-2002	21	16 May	-0.489	$p < 0.001$
			Great tits *Parus major*				
LH	Lanžhot	48° 40' N 16° 56' E	1961-2007	—	—	-0.15	$CI = 99\%$
LN	Lednice	48° 48' N 16° 46' E	1961-2007	—	—	-0.17	$CI = 99\%$
VN	Vranovice	48° 56' N 16° 35' E	1961-2007	—	—	-0.15	$CI = 99\%$
HK	Horka	49° 39' N 17° 11' E	1961-2007	—	—	-0.14	$CI = 95\%$
HP	Hutsepot	51° 01' N 03° 70' E	1979-1998	20	—	-0.509	$p = 0.025$
BW	Boswachter	51° 27' N 03° 70' E	1979-1998	20	—	-0.667	$p = 0.003$
LB	Liesbos	51° 35' N 04° 40' E	1979-1998	20	—	-0.117	$p = 0.383$
OH	Oosterhout	51° 55' N 05° 50' E	1979-1998	20	—	-0.101	$p = 0.626$
WB	Warnsborn	52° 05' N 05° 50' E	1979-1998	20	—	-0.404	$p = 0.017$
HV	Hoge Veluwe	52° 05' N 05° 50' E	1979-1998	20	—	-0.105	$p = 0.540$
VL	Vlieland	53° 18' N 05° 00' E	1979-1998	20	—	-0.200	$p = 0.227$

Note: If the p values are not given, the results are significant under the indicated confident level.

2.2. Data analysis
Past studies have used linear regression to find the slope of MLD versus year in each study site. In this study, theses slopes were categorized into four charts according to species and study area in order to make comparisons between differential long-term trends of MLD. In each chart, the study sites were ranked from smallest to largest in terms of the number of days advanced per year. Subsequently, t tests were used to test whether changes in species' MLD were significantly different from zero in each study area. Then, two-tailed heteroscedastic t tests were used to find whether different species' changes in MLD were significantly different from each other.

3. Results and discussions

3.1. Differential response of pied flycatchers and great tits
On average, pied flycatchers advanced their MLDs by 0.387 days per year (SD = 0.10; $t_3 = 7.49$, $p < 0.01$) across study sites in Central Europe. Great tits advanced their MLDs by an average of 0.153 days per year (SD = 0.01; $t_3 = 24.2$, $p < 0.01$). Similarly, pied flycatchers advanced MLDs by an average of 0.382 days per year (SD = 0.08; $t_6 = 13.1$, $p < 0.01$) in Western Europe, and the great tits by 0.3 days per year (SD = 0.23; $t_6 = 3.50$, $p < 0.02$).
Through comparison of these two species, some general trends were found. In Central Europe, pied flycatchers advanced their laying dates at a significantly faster rate than great tits, as shown in figure 1, which was corroborated by a two-tailed heteroscedastic t test ($t_6 = 4.5$, $p = 0.019$).
Alternatively, long-term trends of MLDs between these two species vary from site to site in Western Europe, as shown in figure 2. For example, the pied flycatchers advanced their laying dates faster than did great tits by an average of 0.361 days per year in Hoge Veluwe (HV) populations, yet the trend reverses nearby in the Warnsborn (WB) populations. This variability in Western Europe can be largely attributed to the varied patterns on a small spatial scale exhibited by the MLDs of great tit populations. It was suggested that variations in microclimate or vegetation and the great tit's life-history, instead of the temperature, might account for the highly varied pattern in laying date [2]. The timing of egg laying for female pied flycatchers, however, is restricted by the timing of arrival at breeding grounds, which is likely to be determined by both the endogenous programme of migration and the environmental conditions on the wintering grounds in Africa. These differences between the two species may explain why great tits advance egg laying faster than pied flycatchers in some sites in Western Europe, whereas not in others.

3.2. Implications on competition for nesting holes
In general, the resident great tits occupy the nest holes for quite some time in advance of the migratory flycatchers. The pied flycatchers then seize nesting holes from great tits in two ways: intense attacks against a great tit when it flies to and from the nest, resulting in nest desertion by the tit; or rapid nest-building in a hole already occupied by a tit when the tit is foraging during the day [14]. The flycatcher's chances of wresting nest holes successfully are likely to decline as the nesting period of great tits progresses. Once the incubation period of great tits starts, takeover attempts are often fatal for pied flycatchers. Therefore, when the interval between the onset of the breeding period between the two species is short, the flycatcher is at an advantage.
Differential advancement in times of egg-laying of the two species has important implications on their competition for nest holes. In Central Europe and few sites in Western Europe (e.g., HV), pied flycatchers are likely to be favored because the interval between the onset of their breeding periods is reduced as a result of pied flycatchers' faster advancement. Conversely, in other sites in Western Europe

(e.g., WB), the flycatchers are disadvantaged because they fail to advance laying dates at a similar pace compared to the great tits. This result agrees with an empirical study of a Dutch population, in which Samplonius and Both find increased pied flycatcher fatalities in great tit nest boxes because tits adjusted egg laying phenology significantly more than pied flycatchers [18].

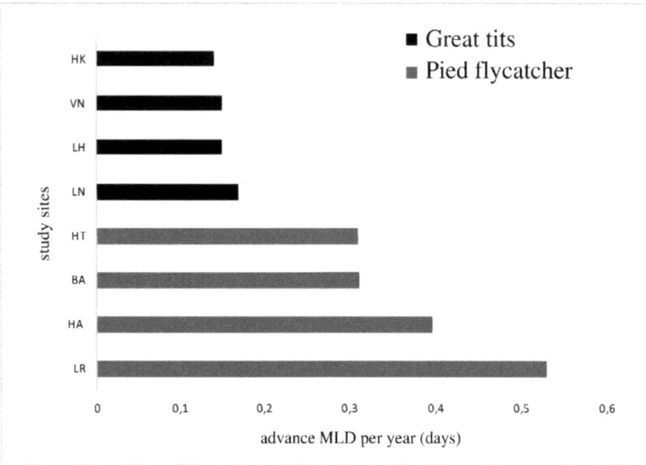

Figure 1. Differential advancement of MLD per year in great tits *Parus major* and pied flycatcher *Ficedula hypoleuca* in Central Europe

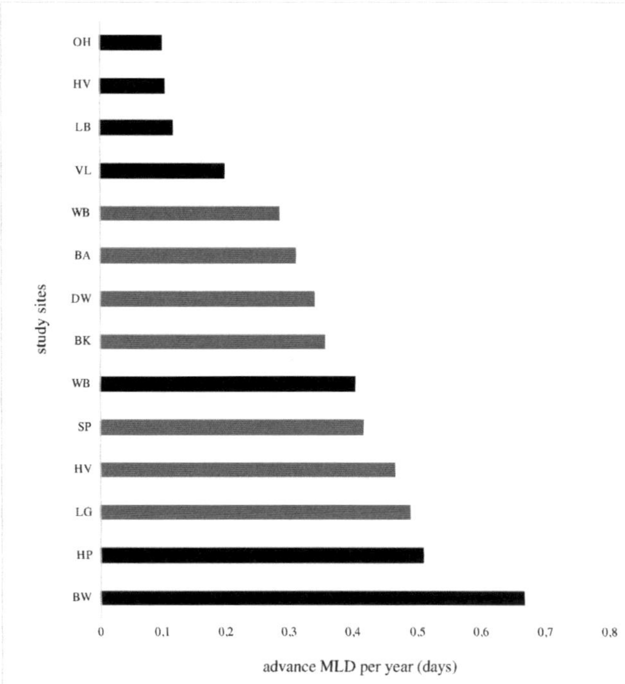

Figure 2. Differential advancement of MLD per year in great tits *Parus major* and pied flycatcher *Ficedula* hypoleuca in Western Europe

Note: In Fig. 1 and 2, black bars represent great tit populations and gray bars are from pied flycatcher populations. The study sites are indicated (Table 1).

3.3. Differential response of pied flycatchers and collared flycatchers

Collared flycatchers advanced their MLDs by an average of 0.19 days per year in Central Europe (SD = 0.02; $t_5 = 28.9$, $p < 0.01$). In approximately the same geographical region, pied flycatchers advanced their laying dates consistently and significantly faster compared to collared flycatchers, as shown in figure 3, which was confirmed by a two-tailed heteroscedastic *t* test ($t_8 = 3.78$, $p = 0.031$).

In Northern Europe, collared flycatchers advanced egg laying by 0.284 days per year on Gotland. In contrast, all surveyed pied flycatcher populations responded slower in study sites near Gotland ($t_2 = 2.97$, $p = 0.097$), with two of the three populations delaying egg laying dates, as shown in figure 4. One possible explanation of the lack of response in pied flycatcher populations is that Northern Europe did not experience significant warming during the study period from 1980 to 2000; therefore, an advancement in laying dates was unwarranted.

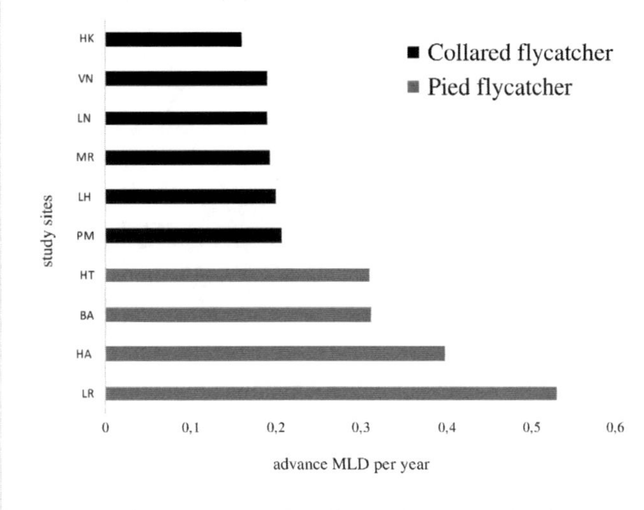

Figure 3. Differential advancement of MLD per year in collared flycatcher *Ficedula albicollis* and pied flycatcher *Ficedula hypoleuca* in Central Europe.

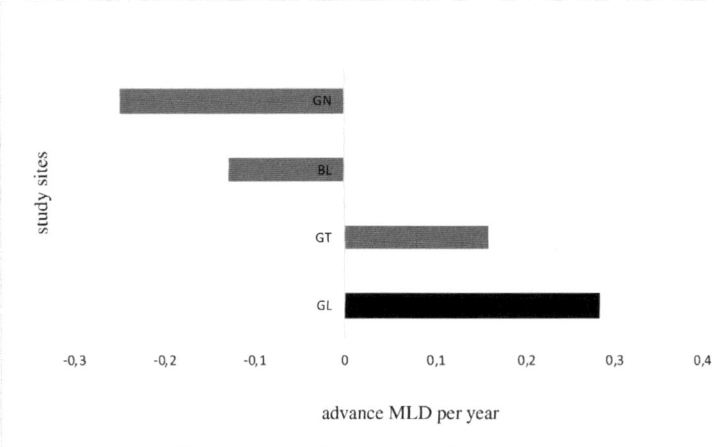

Figure 4. Differential advancement of MLD per year in collared flycatcher *Ficedula albicollis* and pied flycatcher *Ficedula hypoleuca* in Northern Europe.

Note: In Fig. 3 and 4, black bars represent collared flycatcher populations and gray bars are from pied flycatcher populations. The study sites are indicated (Table 1).

3.4. Implications on hybridization

Pied flycatchers and collared flycatchers potentially shared a smaller overlap in breeding dates in Northern Europe as a result of asymmetrical response to climate change. Since the collared flycatcher started egg laying relatively earlier than the pied flycatcher in sympatric breeding grounds in Northern Europe [19], their laying dates could potentially continue to diverge, considering the result that collared flycatcher advanced their MLDs at a faster rate per year. This trend provides both costs and benefits to the pied flycatchers. On one hand, the divergence in the timing of egg laying may reduce the temporal coincidence of mate choice between the two closely related species [20], hence decreasing the occurrence of hybridization. Less frequency of hybridization benefits both species mainly because hybrid individuals hatch fewer eggs and therefore fledge fewer offspring. On the other hand, both clutch size and fledgling success in flycatchers decline as the breeding season progresses [15]. Since pied flycatchers are projected to breed relatively later than collared flycatchers, their overall reproductive success may be negatively impacted due to a later start.

Unexpectedly, the two flycatchers potentially shared a greater overlap in breeding dates in Central Europe, as the laying dates of the two species converged in this region, where the pied flycatchers advanced their MLDs faster. It is perhaps noteworthy that a comparison between the pied flycatcher population in southern Germany (Lahr (LR)) and collared flycatcher populations in the Czech Republic (LH, LN, VN, and HK) revealed that the difference in mean laying dates diminished by several days from 1980 to 2000 due to asymmetric responses between the two species. As a result, the frequency of hybridization is expected to increase in Central Europe.

4. Conclusions

Based on the analysis and discussion, pied flycatchers may be favored in the competition with great tits in Central Europe and certain study sites in Western Europe because they advanced laying dates at a faster rate, whereas this result reverses in other sites in Western Europe. Additionally, the mean laying dates of pied flycatchers and collared flycatchers diverged in Northern Europe and converged in Central Europe due to asymmetrical responses to climate change. Therefore, it is expected that the frequency of hybridization between the two species may decline in Northern Europe and increase in Central Europe. Overall, these results indicate that the effects of differential changes in phenology on interspecific interactions are varied depending on geographical locations, hence more complex than previously thought.

This study has limitations. Firstly, although data from 26 sites are used, the relatively small sample size and absence of empirical studies prevent the author from making definitive conclusions about the consequences of variable responses on interspecific interactions. Secondly, there still exist understudied areas in Europe that are not accounted for in this study due to the absence of long-term trends. As a result, the generality of the results on a large geographical scale may be negatively affected. Thirdly, this study covers wide geographical regions, hence the validity of the results and implications must be confirmed at a local level. Finally, it remains uncertain whether these interspecific interactions could substantially impact community structure and population dynamics.

In the future, field research should be conducted in order to provide empirical evidence for the effects of differential adjustments of laying dates on interspecific interactions outlined in this study. Furthermore, future studies on climate change and its effects on ecological communities should be replicated in different locations in order to account for geographical variability.

5. References

[1] Crick, H.Q.P. (2004), The impact of climate change on birds. Ibis, 146: 48-56. doi:10.1111/j.1474-919X.2004.00327.x

[2] Visser ME, Adriaensen F, Van Balen JH, et al. Variable responses to large-scale climate change in European Parus populations. Proc Biol Sci. 2003;270(1513):367-372. doi:10.1098/rspb.2002.2244

[3] Both, Christiaan et al. (2006). Pied Flycatchers Ficedula hypoleuca travelling from Africa to breed in Europe: differential effects of winter and migration conditions on breeding date. Ardea. 94.

[4] Jonzen, Niclas, et al. “Rapid Advance of Spring Arrival Dates in Long-Distance Migratory Birds.” Science, vol. 312, no. 5782, 2006, pp. 1959–1961. JSTOR, www.jstor.org/stable/3846444. Accessed 14 June 2020.

[5] Nicola Saino, et al. 2010 Climate warming, ecological mismatch at arrival and population decline in migratory birds. Proc. R. Soc. B.278835–842. http://doi.org/10.1098/rspb.2010.1778

[6] Both, C., Bouwhuis, S., Lessells, C. et al. Climate change and population declines in a long-distance migratory bird. Nature 441, 81–83 (2006). https://doi.org/10.1038/nature04539

[7] Walther, G., Post, E., Convey, P. et al. Ecological responses to recent climate change. Nature 416, 389–395 (2002). https://doi.org/10.1038/416389a

[8] Knudsen, E., et al. (2011), Challenging claims in the study of migratory birds and climate change. Biological Reviews, 86: 928-946. doi:10.1111/j.1469-185X.2011.00179.x

[9] Samplonius, JM, Bartošová, L, Burgess, MD, et al. Phenological sensitivity to climate change is higher in resident than in migrant bird populations among European cavity breeders. Glob Change Biol. 2018; 24: 3780– 3790. https://doi.org/10.1111/gcb.14160

[10] Both Christiaan et al. 2004 Large–scale geographical variation confirms that climate change causes birds to lay earlier. Proc. R. Soc. Lond. B.2711657–1662. http://doi.org/10.1098/rspb.2004.2770

[11] WEIDINGER, KAREL & KRÁL, MIROSLAV. (2007). Climatic effects on arrival and laying dates in a long-distance migrant, the Collared Flycatcher Ficedula albicollis. Ibis. 149. 836 - 847. 10.1111/j.1474-919X.2007.00719.x.

[12] Bauer, Zdenek & Trnka, Miroslav & Bauerová, Jana & Mozny, Martin & Stepanek, Petr & Bartosova, Lenka & Zalud, Zdenek. (2010). Changing climate and the phenological response of great tit and collared flycatcher populations in floodplain forest ecosystems in Central Europe. International journal of biometeorology. 54. 99-111. 10.1007/s00484-009-0259-7.

[13] Helm, Barbara & Van Doren, Benjamin & Hoffmann, Dieter & Hoffmann, Ute. (2019). Evolutionary Response to Climate Change in Migratory Pied Flycatchers. Current Biology. 29. 10.1016/j.cub.2019.08.072.

[14] Slagsvold, T. (1978). Competition between the Great Tit Parus major and the Pied Flycatcher Ficedula hypoleuca: An Experiment. Ornis Scandinavica (Scandinavian Journal of Ornithology), 9(1), 46-50. doi:10.2307/3676138

[15] Alatalo, R., Gustafsson, L., & Lundberg, A. (1982). Hybridization and Breeding Success of Collared and Pied Flycatchers on the Island of Gotland. The Auk, 99(2), 285-291. Retrieved June 15, 2020, from www.jstor.org/stable/4085975

[16] Ouwehand J, Both C. African departure rather than migration speed determines variation in spring arrival in pied flycatchers. J Anim Ecol. 2017;86(1):88-97. doi:10.1111/1365-2656.12599

[17] Harnos, Andrea & Ágh, Nóra & Kovacs, Szilvia & Lang, Zs & Csörgő, T. (2015). Increasing protandry in the spring migration of the Pied Flycatcher (Ficedula hypoleuca) in Central Europe. Journal of Ornithology. 156. 10.1007/s10336-014-1148-3.

[18] Samplonius, Jelmer M., and Christiaan Both. “Climate Change May Affect Fatal Competition between Two Bird Species.” Current Biology, vol. 29, no. 2, 2019, doi:10.1016/j.cub.2018.11.063.

[19] Alatalo, R.V., Eriksson, D., Gustafsson, L. and Lundberg, A. (1990), Hybridization between Pied and Collared Flycatchers—sexual selection and speciation theory. Journal of Evolutionary Biology, 3: 375-389. doi:10.1046/j.1420-9101.1990.3050375.x

[20] Ruegg K, Anderson EC, Slabbekoorn H. Differences in timing of migration and response to sexual signalling drive asymmetric hybridization across a migratory divide. J Evol Biol. 2012;25(9):1741-1750. doi:10.1111/j.1420-9101.2012.02554.x

Acknowledgements
I thank Professor André Dhondt and Yuting Qian for guidance and useful discussions. Camille Dolmont commented on the manuscript. I am further grateful for the evaluations given by the anonymous reviewer.

AUTHOR INDEX